# Sowing Legume Seeds, Reaping Cash

Essegbemon Akpo  •  Christopher O. Ojiewo  •
Lucky O. Omoigui  •  Jean Claude Rubyogo  •
Rajeev K. Varshney

# Sowing Legume Seeds, Reaping Cash

## A Renaissance within Communities in Sub-Saharan Africa

**Photo Credits**
David Okello Kalule (1st cover photo)
Justin Ndichu (2nd cover photo)
Agathe Diama
Eleanor Manyassa

Essegbemon Akpo
International Crops Research Institute for
the Semi-Arid Tropics (ICRISAT)
Nairobi, Kenya

Christopher O. Ojiewo
International Crops Research Institute for
the Semi-Arid Tropics (ICRISAT)
Nairobi, Kenya

Lucky O. Omoigui
International Institute of Tropical
Agriculture (IITA)
Kano, Nigeria

Jean Claude Rubyogo
International Center for Tropical
Agriculture (CIAT)
Arusha, Tanzania

Rajeev K. Varshney
International Crops Research Institute for
the Semi-Arid Tropics (ICRISAT)
Patancheru, India

ISBN 978-981-15-0844-8      ISBN 978-981-15-0845-5   (eBook)
https://doi.org/10.1007/978-981-15-0845-5

This book is an open access publication.

**Abstract**

The smallholder farmers' access to the outputs of the breeding programs' achievements has been very weak and hence calls for effective, more coherent, and well-articulated design of technology and seed delivery systems of food legume crops. The Tropical Legumes projects responded to this need.

This book shares impact stories and testimonies from various value chain actors who were part of the Tropical Legumes (TL) projects over the past 12 years. It presents the experiences of a diversity of actors within the grain legume value chains, with a focus on groundnut and common beans in Tanzania and Uganda, groundnut and cowpea in Nigeria, and groundnut in Ghana. All actors involved shared their feeling of being part of decade-long development project families. National agricultural research institutes, knowledge brokering organizations, NGOs, public seed companies, private seed companies, agro-dealers, individual seed entrepreneurs, farm implement makers, farmer cooperatives, farmer groups, individual farmers, women farmers, middlemen, processors, traders, and consumers were all involved in this experience. This book provides learning opportunities for development workers, technical staff, and project managers. It will also inspire development workers and project managers to share their own experiences for others to learn from.

**Keywords**

Grain legume productivity; Smallholder farmers; Impact stories; Multi-stakeholders; Sub-Saharan Africa; Tropical Legumes projects

# Acknowledgments

*Interviewees in Tanzania, Uganda, Nigeria, and Ghana*

We express our gratitude to different actors who accepted to share their valuable experience of Tropical Legumes projects.

*Impact Story Collection*

We thank the communication people who contributed in gathering information from the field. We acknowledge the contribution of Justin Ndichu, a communication consultant who helped to put together the impact stories in Tanzania; Eleanor Manyasa, a communication Intern at ICRISAT, who gathered the impact stories in Uganda, and Agathe Diama, head of communication, ICRISAT-Mali for collecting impact stories from Nigeria and Ghana.

We greatly acknowledge the valuable contributions of Justin, Eleanor, and Agathe who have been instrumental in gathering the various reported impact stories in this book. Without the collaboration of these colleagues, it could have been challenging to deliver on this book.

*Field Facilitators*

We are thankful to the following people for their field support: Dr. Omari Mponda and team at the Tanzania Agricultural Research Institute-Naliendele (TARI-Naliendele); Michael Kilango and team at TARI-Uyole; Ms. Edith Kadenge and team at TARI-Selian; Dr. David Okello Kalule and team at the National Semi-Arid Resources Research Institute (NaSARRI) in Uganda; Dr. Stanley Nkalubu and team at the National Crops Resources Research Institute (NaCRRI) in Uganda; Dr. Oteng-Frimpong Richard and team at the Ghana Savanah Agricultural Research Institute (SARI) in Ghana; and Prof. Candidus Echewku and team at the Institute of Agricultural Research (IAR) in Nigeria.

*Language Edits and Proofreading*

We acknowledge the contributions of ICRISAT communication team.

The Tropical Legume projects transforming the lives of the under-served majority along the commodity value chains in the drought-prone areas

# Contents

**Essegbemon Akpo** is Seed Systems scientist with expertise in plant production, seed systems, innovation studies, participatory action research, and multi-stakeholder processes. He graduated from Wageningen University, the Netherlands. His 15 years experiences in agricultural research for development covered improved seed access facilitation for smallholder farmers, connection between biophysical, social, and institutional landscapes, participatory technology development, innovation platform support, and capacity building for farmers' organizations. He earned several academic awards that permitted him to pursue graduate studies (Research grant within the Convergence of Science Strengthening Agricultural Innovation Systems (CoS-SIS) in 2008, Research grant as part of the Convergence of Science for better Management of Crops and Soils (CoS) Project DGIS in 2005, Research grant as part of the Cowpea Project, Bénin, NWO-FAO in 2003). Dr. Akpo has authored over 15 publications including peer-reviewed papers, books, and conference papers.

**Chris Ojiewo** graduated from Okayama University, Japan. He is a Senior Scientist in Legume Breeding, Global Coordinator of Tropical Legumes III, HOPEII, and AVISA projects (funded by Bill & Melinda Gates Foundation), Cluster of Activities Leader on Science of Scaling Seed Technologies in the CGIAR Research Program (CRP) on Grain Legumes and Dryland Cereals, and Theme Leader of Seed Systems in the Global Research Program on Genetic Gains, ICRISAT. He has more than 60 peer-reviewed international publications and has delivered numerous oral and posters presentations in numerous international meetings. Dr. Ojiewo's research work focusses on basic, applied, and adaptive research and development activities aimed at raising farm productivity, nutrition, and income security for resource-poor smallholder farm households, especially women and youth in rural and peri-urban semi-arid tropics. With expertise in plant breeding and seed systems, he has done extensive work in the development and dissemination of high-yielding and stress (biotic and abiotic)-resilient varieties of vegetables and legumes with farmer- and market-preferred traits together with accompanying integrated crop management practices and efficient seed and technology dissemination systems. Besides promotion of vegetable-legume-cereal-livestock-based family garden intensification systems, improving productivity and profitability for smallholder farmers, gender equity, youth empowerment, nutrition security, knowledge sharing, and solving the perpetual problem of food and nutritional insecurity of the less privileged in developing countries are core to his sense of purpose.

**Lucky O. Omoigui** graduated from the Ahmadu Bello University, Zaria, Nigeria, in 2010. He is a Seed System specialist at the International Institute of Tropical Agriculture (IITA), Kano Station. He has more than 74 peer-reviewed international publications and has delivered numerous oral and poster presentations in numerous international meetings. Dr. Omoigui was an Associate Professor of Plant Breeding and Genetics at the University of Agriculture Makurdi from 2013 to 2016. He received several awards, among which are Most Outstanding Researcher in the University of Agriculture Makurdi in 2012, Excellent Research contribution to West African Cowpea Consortium in 2015, and Arthur Anderson Honour List Award for Best Graduating Student in 1999.

**Jean Claude Rubyogo** is a CIAT Senior Scientist based in Arusha, Tanzania. He focuses on seed systems research and development, technology transfer, and research product commercialization. For more than 30 years, he has contributed/led several research and development initiatives in various agricultural research areas including participatory crop research (both on station and on farm), testing agricultural technologies, and partnership to commercialize and scale up these proven technologies. For the last 15 years, he has been leading public–private multisectoral and multidisciplinary teams developing and deploying sustainable and impact-oriented bean seed systems to serve more than 27 million of smallholder farmers (58% being women) across several member countries of the Pan Africa Bean Research Alliance. Jean Claude has extensively published several articles, books, and book chapters on bean seed systems and technology transfer.

**Rajeev K. Varshney** is Research Program Director of Genetic Gains and Director of the Center of Excellence in Genomics and Systems Biology at ICRISAT. He has more than 15 years research experience in international agriculture. Before joining ICRISAT in 2005, he worked at IPK-Gatersleben, Germany, for five years. While working at ICRISAT, in his dual appointment for six year, he also served CGIAR Generation Challenge Program as Theme Leader. Varshney is recognized as a leader in applied genomics, genomics-assisted breeding, and translational genomics for agriculture. He has genome sequences of 9 crops including pigeonpea, chickpea, peanut, and pearl millet and several molecular breeding products in chickpea, peanut, and pigeonpea to his credit. Varshney is an elected Fellow of Leopoldina—German National Academy of Sciences, American Association of Advancement in Sciences (USA), the World Academy of Sciences, Crop Sciences Society of America, American Society of Agronomy, and all 4 leading science academies of India: Indian Academy of Sciences, Indian National Science Academy, the National Academy of Sciences, and National Academy of Agricultural Sciences. He provides leadership by serving as member/chair for several committees, editorial boards, funding organizations, and advisory boards in international agriculture research, development, and capacity building.

For decades, the vast majority of smallholder farmers in developing countries, mainly sub-Saharan Africa (SSA) and to some extent South Asia (SA), heavily rely on non-improved and auto-saved variety seed, accounting for about 80% of their material used for planting. Though the seed use figures by farmers vary from one region to another, with West Africa showing the lowest rate of improved seed use (below 20%) and South Asia with much higher rate (up to 70%), the overall situation looks less encouraging. At the same time, substantial breakthroughs have been made by breeding programs, and many more are still in the pipeline. Some of the traits of recently developed varieties have targeted the consumers' demands and farmers' preferences.

The smallholder farmers' access to the outputs of the breeding programs' achievements has been very weak and hence calls for effective, more coherent, and well-articulated design of technology and seed delivery systems of food legume crops. The Tropical Legumes projects answered this historical call.

The Tropical Legumes projects (TL I, II, and III) are an international research and development initiative under the partnership of the International Crops Research Institute for the Semi-Arid Tropics (ICRISAT), the International Center for Tropical Agriculture (CIAT), the International Institute of Tropical Agriculture (IITA), seven African countries (Burkina Faso, Ghana, Mali, Nigeria, Ethiopia, Tanzania, and Uganda) and one Indian state (Uttar Pradesh), and others. It aimed to develop improved cultivars of common bean, cowpea, chickpea, and groundnut and deliver their seed at scale to millions of smallholders. Since the beginning of Tropical Legumes I (TL I) project in 2007, and the subsequent TL II and TL III projects, a lot has been achieved at different levels of the commodity value chain of each focus crop. The impact can be felt from farmers in remote communities who are now smiling all the way to the banks, to the research centers involved, thanks to this major research and development initiative led by ICRISAT.

This publication covers impact stories from an array of actors within the crop value chain, with a focus on groundnut and common beans in Tanzania and Uganda, groundnut and cowpea in Nigeria, and groundnut in Ghana. All actors were given an opportunity to share their perceptions and stories for being part of TL families. These actors involve: National agricultural research institutes; knowledge brokering organizations, e.g., extension services and NGOs; public seed companies; private sector operators, e.g., private seed companies, agro-dealers, and individual seed entrepreneurs; farm implement makers; farmer organizations, e.g., farmer cooperatives, farmer groups, individual farmers, women farmers; and the end-pull investors, e.g., middlemen, processors, traders, and consumers. This document shares records of project achievements through impact stories, testimonies from various value chain actors who benefitted from the TL projects over the past 12 years.

It is our hope that the reader will get inspired reading through great field stories from stakeholders over the 12 years of research and development processes implemented to put smiles on faces of families in farming communities in the dryland areas of focus countries.

## 2.1 Farmers and Farmer Groups in Remote Communities Share Their Benefits from TL Projects' Investments in Tanzania

### 2.1.1 Farmer Sharing Asset Enhancement Through Groundnut Production

Adamu, a groundnut farmer from Maugura village, Masasi, shared his success story for being involved in TL projects (Figs. 2.1 and 2.2). *"I was taught how to grow the seeds, carry out diagnosis, how to store them, among other things. This year, there are some seeds that I have begun putting on the ground so that I can continue conducting research about them. Naliendele Institute gave me about 20 lines and I am working on all of them. In fact, they have not yet been named. I've just planted them in plots; from plot number one to plot number twenty."*

Nyirenda is reaping big from his seed production business. *"Last year, I got 90 bags of groundnut from 4 acres, and I sold 47 bags through Naliendele. I sold to other farmers the remaining 43 bags"* he said. One bag equals to 42 kg.

Nyirenda does not regret his decision to venture into groundnut seed production. *"First, I have six children; two are in secondary school and two are in primary school. I pay their fees from the money I earn in the groundnut business. I have built a good house and bought more land to expand the planting area from the proceeds of the groundnut business as well. Generally, I would say, for me this a self-sufficient business."*

Nyirenda, however, appeals to the government to purchase planters on behalf of the farmers as this will reduce the cost of production and increase profits. He also thinks that if a factory is set up for groundnut value addition, farmers like him will not struggle any longer with lack of market.

*Groundnut have transformed my Life,*—Adamu Nyirenda.

© The Author(s) 2020
E. Akpo et al., *Sowing Legume Seeds, Reaping Cash*,
https://doi.org/10.1007/978-981-15-0845-5_2

**Fig. 2.1** Mr. Adamu Abilah Nyirenda with part of his family outside their house constructed with proceeds from groundnut in Maugura village, Masasi District, Tanzania (Photo: Ndichu J)

**Fig. 2.2** Mr. Adamu Abilah Nyirenda, Maugura village in Masasi District, Tanzania shows bricks he was able to acquire to do more constructions in his homestead, all from groundnut proceeds (Photo: Ndichu J)

In March 2018, we met Adamu Abilah Nyirenda who is now a seed producer at Maugura village in Masasi District. Nyirenda who has been closely working with Naliendele Research Institute says he was fully involved in developing of the *Nachingwea* variety.

## 2.1.2   Farmer Groups Take Up Groundnut Farming as a Business

Among different groups sampled, many took up farming groundnut as a business in different regions of Tanzania. The results are impressive as all groups are now reaping big from this business and they are not turning back. Examples of these groups involve:

- Hekima Farmers' Group located in Nzali, Chamwino District, Central Zone. The group has 13 members, 2 men and 11 women, currently doing seed multiplication.
- Jumatuwa Farmers' Group located in Nzali, Ushetu District, Lake Zone. The group has a total of 18 members, 2 men and 16 women.
- Jipemoyo Group located in Kinanga, Kahama District, Lake Zone. The group has 30 members, 14 men and 16 women.
- Upendo Women Group located in Mnanje, Masasi District, in Southern Zone. The group has 25 women.
- Owe Faraja Group located in Mindola, Bahi District, Central Zone. The group has 15 men.

Apart from farming as groups, all of them encourage individual members to plant in their household farms, thereby increasing productivity and multiplying benefits.

We talked to members of these groups and sampled a few different reactions they had to say.

Ms. Anastazia Thomasi Madeje (Fig. 2.3) is Hekima Famers' group secretary. She still vividly remembers some of the teachings they received through Naliendele

**Fig. 2.3**  Ms. Anastazia Thomasi Madeje; Hekima Famers' group secretary at the Group's farm in Nzali, Chamwino District, Tanzania (Photo: Ndichu J)

Institute. *"Initially, we used Pendo variety but later this year we were given three other varieties by the Chamwino District Council; Naliendele, Mangaka, and Mnanje. I've planted these three according to the dimensions the agricultural officer directed us. Each seed has its bed, with spaces of up to 50 centimetres from one line to the other, 10 inches from one hole to the next and the size of each bed is 15 meters,"* she informed us.

Mr. Andera Shoko Kayanda, Secretary, Jumatuwa Farmers' Group (Fig. 2.4), helps us to compare the old groundnut varieties and the new improved ones. *"There are several differences; for instance, with Pendo, women say while preparing meals it's easy to mix with vegetables, they are soft, white in colour, have a lot of fat content and are sweet,"* says Kayanda.

Mr. Francis Paschal, a member of Jipemoyo Group, says through the proceeds they get from selling groundnut seeds, his group has been able to purchase 1 acre piece of land, which is not enough and they have hired 2 more acres (Figs. 2.5 and 2.6). *"Our lives have improved,"* says Paschal. *"Groundnut grows fast compared to other crops that we grow. We deal with rice, maize, and groundnut. After planting in November, we can harvest groundnut; at that time maize and rice will still in the ground. This enables us to have cash-flow as we wait for maize and rice to harvest"*.

Ms. Judith Msonje Masaka, the Chairperson of Upendo Women Group (Fig. 2.7), has also seen the difference between the old varieties that farmers were used to and the new improved varieties they have embraced. *"These varieties are big in size and do not require much work during the breakthrough. Last year, I personally ploughed one acre and got fifteen sacks which I sold to improve my life at home,"* says Masaka.

Owe Faraja Group in Mindola village, Bahi District, is a relatively youthful group (Fig. 2.8). Mr. Stephano Joseph, the group chairperson, reported that they are currently doing trials. *"In our 1 acre piece of land, we have Mangaka, Naliendele, Mnanje and an old check variety,"* says Mr. Maongezi Mohamedi Hassani, the

**Fig. 2.4** Mr. Andera Shoko Kayanda, Secretary, Jumatuwa Farmers' Group shares their TL stories in Sabasabini Village, Ushetu District, Tanzania (Photo: Ndichu J)

**Fig. 2.5** Members of Jipemoyo Group showing their produce at their go-down in Kinanga, Kahama District, Tanzania (Photo: Ndichu J)

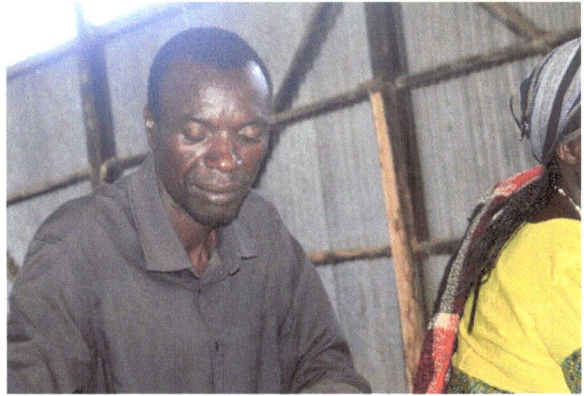

**Fig. 2.6** Mr. Francis Paschal, a member of Jipemoyo Group, Kahama District, Tanzania (Photo: Ndichu J)

group secretary. Faraja says they got interested in farming groundnut after seeing how well other farmers in Mindola were doing after planting groundnut. *"We got seeds from Naliendele and our expectations are to continue growing this crop. We hope to harvest the seeds and plough them back to expand our acreage."*

Our farmer organisation has increased its production of groundnut from 50 kg to about 300 kg currently. *"The value of the organisation has definitely risen. If you look at our equity and the infrastructure, putting everything together, the asset base has risen to over Tsh 70 million. I no longer rely on buying and selling, I am producing and selling. Then we have already expanded to other villages where we have other farms, for instance in Dodoma. There is still greater room for growth as long as we are in legume production,"* Mr. Chaula discloses to us.

The organisation's future looks bright; Chaula (Fig. 2.9) is currently writing a proposal to look for funds to put up a center pivot irrigation system infrastructure on his farms.

**Fig. 2.7** Members of Upendo Women Group learning from researcher Charles Mkandawile at their farm in Mnanje, Masasi District, Tanzania (Photo: Ndichu J)

**Fig. 2.8** Mr. Stephano Joseph (third right) with other members of Owe Faraja Group introducing visitors to their farm in Mindola village, Bahi District, with Mr. Anthony Sahali (first right) and Mr. Joel Mpayo (second right), Tanzania (Photo: Ndichu J)

**Fig. 2.9**  Mr. Aithan Chaula showing DASPA farm in Dodoma, Tanzania (Photo: Ndichu J)

### 2.1.3   Informed Decision-Making for Groundnut Production

Apart from those farmers who farm groundnut for seeds, these farmers grow groundnut for subsistence use. They are basically grain framers. The 30-year-old Bushiri Rashidi Selemani, from Mpeta, Masasi District, whom we also met in March 2018, told us that he previously planted *Pendo* variety in his 1 acre piece of land and harvested 15 sacks of groundnut, despite planting late. *"If I plant in good time and the rain is abundant, I will reap 20-25 bags for one acre."*

Although he grows groundnut for food, Selemani said that with such harvests, he is selling the remainder to other farmers in his village. *"I have been able to buy a bicycle, when I get a family of my own, I will take my kids to private school through this income, and still have money in my pocket. My life has surely changed,"* he stated, smiling.

Selemani (Fig. 2.10) would not let us go before passing an appeal to his government. *"My call to the government is to take care of groundnut farmers the same way it has done for cashew nut farmers and get us planter machines, even if it's through loans so that we are able to grow in time. You know we can only harvest more if we plant in time."*

### 2.1.4   Improved Groundnut Varieties Planted at the Dispensary Farm to Teach the Nutritional Benefit to Communities

In late March 2018, we went looking for people who have benefited through efforts of TL projects especially through groundnut in one way or another. We came across two institutions that have taken up farming of groundnut and are playing a significant role in knowledge creation and awareness.

**Fig. 2.10** Mr. Bushiri Rashidi Selemani shows his bicycle in Mpeta, Masasi District, Tanzania (Photo: Ndichu J)

**Fig. 2.11** Dr. Steve Julius (right) with Mr. Anthony Sahali at Ilindi Dispensary's farm in Ilindi, Bahi District (Photo: Ndichu J)

First is Ilindi Dispensary located in Ilindi, Bahi District (Figs. 2.11 and 2.12). Dr. Steve Julius, the doctor in charge, says they took up an initiative early 2018, through the help of researchers in Naliendele to plant groundnut at the dispensary farm and use that to teach patients about the benefits of taking groundnut as food. *"The nutritional benefit of groundnut in the baby's body is the protein value. Groundnut in the porridge help with the growth of the child. Also, pregnant women need protein to avoid giving birth to underweight children."*

**Fig. 2.12**  Establishment of Ilindi Dispensary in Ilindi, Bahi District (Photo: Ndichu J)

Through these demos, Julius says that the community has taken up the lessons seriously and implemented the knowledge in their household farms. *"After giving them training, there are about fifteen mothers who went ahead and planted the seeds at their home and they came to show us the results,"* he adds.

### 2.1.5   Agricultural School Facilitating Community Access to Improved Varieties of Groundnut

In Mnanje, Nanyumbu District, we met Mr. Ajili Mkero who is a teacher at Nanyumbu Primary School. He reported that the school has been growing groundnut since 2015 (Fig. 2.13). *"Naliendele has been bringing these seeds to us every year and we can say that the collaboration has been good. I was fascinated by a group of women who were growing the new improved varieties, I started growing them myself, and I later introduced to my school."*

One would be interested to know how the school goes about growing groundnut. Mr. Mkero says, *"We teach the children how to plant groundnut in school and the groundnut become food for them. The children also get nutrients they deserve."*

Other than getting food from the groundnut, the school has been able to benefit from selling the seeds. *"When we sell groundnut as a school, we direct those funds to 'Elimu ya Kujitegemea' (self-reliance) department."*

We can, therefore, say that relatively TL projects have been able to impact in a big way the lives of many people in Tanzania. In the local Swahili language, Tanzanians say, *"Mgeni njoo mwenyeji apone,"* meaning let the guest come so that the host may benefit. That is exactly what the TL projects did when they were rolled out in Tanzania.

**Fig. 2.13** Mr. Ajili Mkero between his students of Nanyumbu Primary School in Mnanje, Nanyumbu District at the school's farm, far right is Extension Officer, Ms. Leodina Ernest Mpagama (Photo: Ndichu J)

## 2.2 Seed Companies and Agro-Dealers Got Interested in Groundnut Seed Business

### 2.2.1 Seed Companies Venture into Groundnut Business

Prior to the TL projects in Tanzania, private seed companies were shy from selling groundnut seeds, either due to lack of market or lack of interest to do so. Things have since changed. Aithan Chaula, apart from being the District Agricultural Officer in Chamwino District, is also the manager of Dodoma Agricultural Seed Production Association (DASPA). *"We ventured in the production of certified seeds because for over 10 years back, there were no such seeds here. We knew that many farmers were using their own recycled local seeds, so we introduced groundnut into the business, and by next year most of the agro-dealers will have more of our groundnut to sell,"* Mr. Chaula reported.

Mr. Chaula, through his association, is currently selling his products to over 10,000 farmers within and outside Dodoma. *"We sell in packs of one and two kilograms directly to farmers but in the last two to three years, we have had organizations that are buying these seeds from us and distributing to farmers."*

### 2.2.2 Agro-Dealers Have Improved Their Business in Tunduru

We met Ms. Zuwena Hamisi Chipangula (Fig. 2.14) at her company' shop, Tunduru Agro-Dealer in Tunduru District within the Southern Zone, in March 2018. The company began operations in 2002 and has been selling improved groundnut

**Fig. 2.14**  Ms. Zuwena Hamisi Chipangula working at her shop; Tunduru Agro-Dealer in Tunduru District, Tanzania (Photo: Ndichu J)

varieties since 2013. *"We deal with the sale and supply of agricultural products in southern Tanzania. The varieties of groundnut we are selling are Pendo, Mangaka, Naliendele, Mnanje, and Nachingwea,"* she says.

Chipangula has now familiarized herself with these new varieties and she helps us identify traits her customers like from these varieties. *"Pendo is a variety that grows in a record 90 to 110 days. It has the capacity to give 1500 kg per hectare and can cope with diseases. Mangaka also grows within 90 to 110 days. It can give 1600 kg per hectare and provides three seeds for one pod. Naliendele also matures within 90 to 100 days, and it has the capacity to give 1000 kg per hectare. Mnanje, on the other hand, grows within four months and can produce up to 1000 kg per hectare. Lastly, Nachingwea grows in 110 days and it can produce 1000 kg per hectare."*

Since she started selling the improved varieties, Chipangula says her business has grown by leaps and bounds. *"We pack from half a kilo onwards, and so far, we have close to 1000 farmers who are our loyal customers. We sell four tons of seeds each year,"* she says.

Chipangula says she has been able to further her education through this business. *"My business has grown because more people have gotten information about these new varieties and their characteristics and have embraced them."*

## 2.3    Groundnut Processors and Traders Boosting Their Business and Making More Cash

### 2.3.1    Processors in Tanzania Now Processing Cash

When an innovation comes, it does not just benefit its primary users; you may be surprised that others will take it up and use it more innovatively to make a living out of it. That was the case with Mr. John Julius Bakari (Fig. 2.15), who owns Temnar Company Limited, a company that makes vegetable oil. Temnar Ltd. is located in Masasi District.

He says when the new varieties of groundnut were introduced in Tanzania, he did not hesitate to engage with the researchers, to benefit from varieties with high oil content, through value addition. *"Naliendele Institute has assisted us by providing groundnut that are rich in oil content, which is what we are interested in. When you look at the indigenous varieties, you can't compare with these improved ones. So, their technology has helped us a lot,"* Bakari reported.

He has mostly been selling his products to retailers as he says many wholesalers have not known about these new products, but he hopes with time, more people will embrace his oil products.

Bakari is a man who is full of gratitude. He says this innovation came to benefit people like him. He stated, *"My Company has benefited a lot; I have seen its income rise since I now spend less on production while the output is high."*

**Fig. 2.15**  Temnar company installation (Photo: Ndichu J)

### 2.3.2   New Business Opportunities Through Purchase and Sale of Groundnut in Kahama

After talking to Mr. Samson Sumuni (Figs. 2.16–2.18) whom we saw earlier, as an agricultural extension officer, he mentioned that in Kahama Mjini, several business-men have taken up the business of buying, shelling, and selling groundnut. We, therefore, sought to speak to one such businessman, and he took us to Mr. Simioni Edward, the owner and manager of Mungula Shelling Plant and depot at Mungula, in Kahama. This was in March 2018.

Simioni reported that in 2015 he saw an opportunity with the readily available groundnut from farmers. He now buys the produce from local farmers and exports the same to the neighboring Uganda and Rwanda. With a shelling machine and godown in his premises, Simioni can shell and store groundnut to sell in large quantities. *"My machine has the ability to shell 200 sacks a day, while my warehouse can take about 60 tons when full."* His business, he says, cost him about Tsh 20,000,000 (about USD 8700) to set up.

On a good day, Simioni can buy groundnut from around 60 farmers each day. *"For a day I can buy three or four tons and on a good day, I am able to buy up to twenty tons."*

He has created employment through his business. *"I have employed four workers on a permanent basis, but those who benefit directly are more. We have three more people who get up to Tsh. 20,000 (USD 9) daily by helping us to load and offload the goods."*

Simioni said there are about three other plants like his in Mungula village alone! This speaks volumes about the impact of TL projects in Tanzania as a whole.

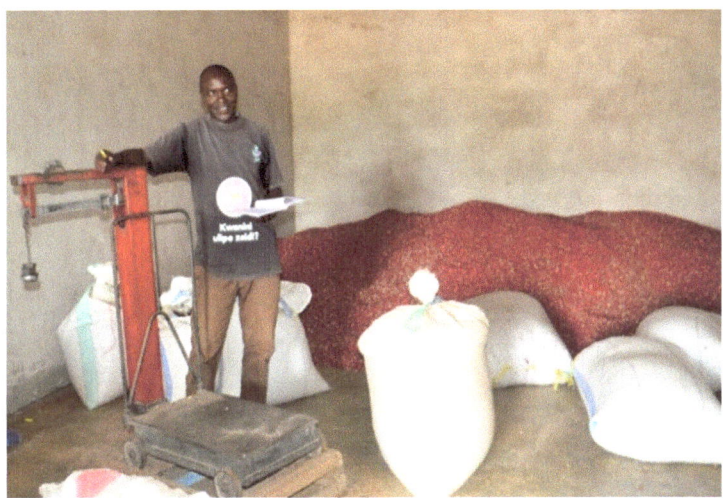

**Fig. 2.16**   Mr. Simioni Edward in his godown in Mungula, Kahama, Tanzania (Photo: Ndichu J)

**Fig. 2.17** Mr. Simioni Edward (extreme right) at his Mungula Shelling Plant in Mungula, Kahama, Tanzania (Photo: Ndichu J)

**Fig. 2.18** Workers at Mungula Shelling Plant and depot at Mungula, in Kahama, Tanzania (Photo: Ndichu J)

In Ushetu, Manguzi (Fig. 2.19) reported that, due to the high productivity of groundnut, shelling machines can be heard all over the town. *"Groundnut prices do not fluctuate like maize, giving the farmer a fairly stable income through the year,"* said Manguzi.

**Fig. 2.19** Mr. Aron
Manguzi walking with us
in Ushetu District,
Tanzania (Photo: Ndichu J)

**Fig. 2.20** Mr. Athanas Minja (right), with his colleague Happy Daudi, inspecting groundnut varieties at the Naliendele Institute's Crop Museum, Mtwara, Tanzania (Photo: Ndichu J)

## 2.4    Researchers Share Their Views of Groundnut Breeding and Research Efficiencies Enhancement in Tanzania

### 2.4.1    The Breeder Who Is Glad to Have Been Alive When the TL Projects Set in Tanzania

Mr. Athanas Joseph Minja (Fig. 2.20, right) is currently managing groundnut breeding under TL projects and working with farmer research groups to popularize groundnut varieties and crosses; developing groundnut varieties that are resistant to biotic and abiotic stresses like rosette disease and drought; and creating awareness on the issue of aflatoxin and the improved varieties.

Minja has been involved in the Tropical Legumes projects since 2007; we talked to him in March 2018.

*We have been receiving the new germplasm from ICRISAT Malawi and ICRISAT India and we are using them to improve our released varieties in the crossing so that we can integrate those traits of drought tolerance and oil content. From ICRISAT Malawi we also received germplasm that are rust resistant,* says Minja. He adds, *We evaluate them, do a preliminary evaluation. From there, we do participatory variety selection (PVS) with the farmers to compare factors like yield and other traits. We then involve Tanzania Official Seed Certification Institute (TOSCI) and release after passing through all those stages.*

The research institute has been able to release 17 varieties in total since 2007, 13 of which were released under TL projects. *"Currently we have five set of trials; close to 1000 lines. The size of the nursery has been increasing, we are evaluating close to 1000 nurseries in Naliendele."*

*Recently we were able to release Naliendele 2016, Mtwaranut 2016 and Tazanut 2015. These varieties are high yielding, more than 1 t/ha, and are preferred by the farmers and consumers in the market. This is because they take a short duration, therefore preferable in agro-ecological zones that have short rains. We also have those that fit heavy rain ecological zones like Southern Highlands. In such places, we advise farmers to use varieties that take medium duration like Mnanje 2009, or Naliendele 2016.*

Minja says they have been working with different stakeholders, to multiply the improved new varieties, namely, Non-Governmental Organizations like World Vision and Agha Khan Foundation. In Mbeya district, which is found in the Southern Highlands, they have also engaged the Agricultural Seed Agency (ASA), a Public Seed Multiplier, and other private seed multiplication companies like Suba Agro, Agri-Seeds, and Meru Agro.

He says that the demand for these improved varieties has been growing high, due to popularization through the media. *"We have engaged national television broadcasters like Tanzania Broadcasting Corporation (TBC), Azam TV, Star TV, and ITV. We also have engaged radio stations like TBC Radio, and newspapers like The Guardian and Daily News, to cover this information. Recently we got agricultural field officers from Katavi who said that they want to cover more than 20 acres of land with these new varieties since the ones they have been using have been suffering from rosette,"* a visibly cheerful Minja said.

Gender aspects have also been considered when releasing the varieties. *"Women prefer the soft pod for ease shelling. They also prefer varieties with high oil content for relish making. Others would be colour; and you find that colour is sometimes co-related with things like iron nutrients. Some prefer red, others tan colour. Based on that information we can know during the crossing, what they prefer in the market. Traders prefer varieties that are big in size, while processors look at the oil content."*

In terms of resources, Minja says TL projects have been very crucial. *"For groundnut, we used to produce 300kgs/ha, but with the use of this new technology and with proper spacing and taking care of the field, we have been able to realize 1 t/ha. Productivity has improved and now more farmers are planting groundnut especially the improved varieties. Farmers are now using proceeds they get from groundnut to manage the perianal crops, like cashew nuts."*

Happy Daudi inspecting her breeder varieties at
Naliendele Institute in Mtwara, March 2018

**Fig. 2.21** Ms. Happy Daudi, a breeder who is also based in TARI-Naliendele in Mtwara, is undertaking her PhD under TL projects sponsorship, Tanzania (Photo: Ndichu J)

### 2.4.2   Happy, a Breeder at TARI-Naliendele, Is Now a Happier Breeder

*I am going to study groundnut breeding, especially on rust disease. As we know rust disease is a problem in coastal areas, where temperature and humidity are high,* she said.
*I currently have 120 breeding lines, 109 of which I got from ICRISAT Malawi, while others are the varieties that we released from Naliendele. I also have some old varieties that I collected from Shinyanga, Dodoma and here in the Southern Zone. My nursery is about 120 meters.*

Daudi (Fig. 2.21) says the TL projects have been instrumental in her research. *"I have benefited a lot as a student under TL through capacity building; I attended training on how to identify the disease and the impact they have through their breeders. We also received some germplasm from ICRISAT."* This she says has strengthened their breeding program. *"At the end, we will be more independent than dependent, because we have a lot of resources, like the germplasm."*

## 2.5   TL Projects' Investments in Research Infrastructure of the National Agricultural Research Systems

### 2.5.1   Irrigation Facilities Fully Installed at the Tanzania Agricultural Research Institute-Naliendele, Tanzania

Daudi is thankful for the support that breeders at their research center got from the TL projects. *"Other resources we have received are like this irrigation facility, this will help us because even for my study I will use irrigation."* (Fig. 2.22)

**Fig. 2.22** TL III built irrigation facility at TARI-Naliendele, Mtwara, Tanzania (Photo: Ndichu J)

On the varieties she is working on in her study, Daudi had this to say, *"For my PhD, I selected a few varieties, I have around 6 improved varieties, the local ones, and new lines that have not been released. In the end, I will come up with my variety which is resistant to diseases. I have not been able to release any variety, because I am in my second year of study, so I am still going to the trials."*

From her own experience, she has used varieties like *Pendo* which she says is an improved variety although there are new developments. *"Farmers prefer it a lot, but it has its own challenges. It is susceptible to all foliar diseases like rosette, rust, and leaf spot, which is the main challenge. But the new ones are resistant to diseases, are high yielding, and have high oil content."* She adds that, more awareness on the newer varieties needs to be done so that farmers can embrace and reap from them.

### 2.5.2  The TL Projects Have Catalyzed Our Work of Research, at TARI-Naliendele, Mtwara

Dr. Omari Mponda (Fig. 2.23), the acting Southern zone regional director, TL projects principal collaborator in Tanzania, and a Seed System Scientist, is grateful that The TL projects happened in his country. We met him in his office in March 2018. *"Tropical Legumes projects have helped because most of the scientists had already been trained in farming systems research approaches. The training focused on the need to involve the client in the technology development. Therefore, we have been very successful in disseminating our groundnut varieties through those participatory approaches. We can now say that in Africa, Tanzania is currently second after Nigeria which is producing 3 million metric tons while in Tanzania, we are producing about 2 million metric tons. So, the graph on the production of groundnut has*

**Fig. 2.23** Dr. Mponda (extreme left) takes us round the Naliendele Institute's Crop Museum, Tanzania (Photo: Ndichu J)

*been on the increase and growing very fast from 2007 when we started till now. I think we have done a lot of things,"* says an upbeat Mponda.

He reported, *"Within groundnut, we have several objectives, but the main ones are new varieties that are resistant to biotic and abiotic stresses, including some agronomic recommendations that will increase productivity at the farm level."*

Dr. Mponda adds, *"We have a seed roadmap, to make sure that over the 4 years of support from Tropical Legumes, we produce about 8000 metric tons of groundnut certified seeds in Tanzania. Therefore, we have been trying to reach the target over the years, and each year we produce something depending on how many farmers we have reached. But I think we have done very well in groundnut seed production in Tanzania."*

TL projects impact cannot be simplified better than this: *"Initially the average productivity for groundnut was only 300 kilogram per hectare, but now it's almost 1 t/ha because of the adoption of the new improved varieties which have improved productivity,"* Mponda breaks down.

Dr. Mponda compares what his research institute was able to achieve over the years with what they have achieved in about 10 years of TL projects in Tanzania. *"From our research, we have at least 13 varieties of groundnut that we have released since 1978 with previous programs. Now with Tropical Legumes projects in general that started in 2007, I think we have been able to release 11 varieties because the first variety was released in 2009, then 3 varieties were released in 2015, and now this year we have released 3 varieties adding up to the 11 existing varieties. These varieties command high yielding ability, they are resistant to rosette and some of them are drought tolerant. Also, we develop varieties depending on their demand. The Tanzania market requires a conventional market, so we develop varieties that are large-sized. We are also developing varieties that are high oil content, so we can have some groundnut vegetable oil from it."*

We ask about the varieties that have been released and how they are faring in the market. *"Of the most popular varieties, one is Pendo that was released before TL III and this is our own variety which I think the adoption rate is something about 60%. With Tropical Legumes support, we have another variety called Mnanje that is also commanding high popularity. The other one is Mangaka and this is one of the varieties that was developed from ICRISAT, we tested it in our elite variety trial and eventually, we released it. It is also high yielding and mostly preferred in the market. The Tanzania market requires varieties that are easy to shell. Mangaka is good for them because it is easy to crack, so they don't use a lot of force, which is why it is highly preferred. Another variety that is common in Tanzania is Naliendele 2009. This one is a drought tolerant crop, early maturing and here in southern Tanzania it is largely grown in Masasi which is a drought-prone area. They receive very little rain, and I think it is also good for people in the arid areas like Dodoma and Singida. This one will be highly accepted by those farmers."*

Perhaps, a point is easily driven home when it's heard from the key stakeholders themselves. Dr. Mponda narrates a story of a farmer who was pleasantly surprised at how much she could reap from groundnut. *"I had a call from one of the farmers, Angelina in Singida. She told me that the Naliendele 2009 variety I gave her has done wonders. It is very small in size but when you uproot it, it has so many groundnut. She was very happy!"*

The Research Institute has done pretty well in popularizing the varieties with the support of TL, according to Dr. Mponda. *"At Naliendele, we have the Zonal Information Extension Liaison Unit. This one links to research and extension as it is the one responsible for organizing farmer field days, agricultural shows, producing leaflets and responsible for coming up with press releases to link with the media. Every year, we host media personnel either from Tanzania Broadcasting Corporation (TBC) or ITV whereby we send them to our project sites, they talk to farmers and get to see what the farmers are getting from the TL projects, and the challenges they may be facing. So, we broadcast our activities through radios and televisions. We could have 10 to 20 programs in a year."*

The greatest challenge as Dr. Mponda puts, however, is the lack of enough market for what the farmers are now able to produce. *"In Masasi, some farmers in one year failed to sell their groundnut because there were no traders to buy the products, but if the market is well known, they will be able to sell. We therefore need improvement in that area. Otherwise, we have done a lot in convincing farmers to adopt the improved varieties, and productivity increased immensely."* He adds, *"What is actually hindering further increase is lack of a marketing board for groundnut. I remember in 1970s and 1980s, we had general agricultural products for exports marketing body, and it was helping farmers in marketing their produce outside the country; there was always a market for all produce and the farmer knew where to sell. After that era, people have been producing but with nowhere to sell. We now need a groundnut' marketing board so that farmers can get encouraged to produce more. Currently the available buyers collude and buy at very poor prices. My appeal to the government of Tanzania is to assist these farmers by marketing their products."*

## 2.6   Sensitivity Towards Gender Equality and Equity

### 2.6.1   TL Projects Attribute Success to Among Other Things, Sensitivity Towards Gender-Related Issues

In a project of this magnitude, one cannot forget to consider the question of gender if the project is to release success. Professor Joseph Hella is a gender and social scientist at Sokoine University of Agriculture and has been working on the TL projects since 2015. He narrates his experience working with farmers to promote the adoption of improved groundnut varieties in Tanzania.

*When you look at groundnut, it is a woman's job in that it's labour intensive. It requires more work. Since it's a highly marketable product, the men come in at marketing stage. So basically, it's a woman's crop in the garden and the men take over after harvesting.*

#### 2.6.1.1  Professor Joseph Hella in Morogoro, Tanzania, in March 2018
*The new improved legumes uptake has been impressive among women. The women have seen the benefits especially now when the rains have been short but there is a lot of potential in bridging the gender yield gap,* says Prof. Hella.

On the question of technology use among men and women, Hella (Fig. 2.24) had this to say, *"The use of technology is really low. We collected this data from 900*

**Fig. 2.24**  Professor Hella with a groundnut trader in Morogoro, Tanzania (Photo: Ndichu J)

*farmers and their productivity is really low if you compare their potential yield and actual yield. In terms of percentage, the yield from the women is really high compared to the men while the marketing percentage from the men is high compared to the women."*

There has been the use of innovation too. *"They implemented methods to ease the labor input directed towards groundnut growing. The use of these good seeds has improved their yield and they tend to grow fast so it's something positive. The drawback is the number of rains that come. If the rains are heavy, then the groundnut sometimes may not withstand. Innovations in terms of marketing, TL has installed a lot of knowledge which has seen farmers getting involved in variety choosing. Life changing decisions on farming groundnut has seen farmers meet their basic needs. The farmers have also developed group marketing seeing those targeting specific markets bringing in a value chain,"* he reported.

Professor Hella has witnessed first-hand, the impact that adoption of the new varieties has had among the farmers. *"The lives of farmers have changed since they are getting good harvests."* There has not been without challenges though. *"The problem they have is getting seeds. This can be corrected through knowledge sharing and communication. Gender wise we are pegged at 20 percent in reaching out to farmers who are using improved seeds. We don't have good machines logged towards groundnut. With groundnut, we have a lot of variety but what we lack is the multiplication of seeds. Few farmers know about these seeds."*

## 2.7    TL Projects Extended Research Capacity Building Beyond Focal Research Centers Across Tanzania

Ms. Felista Joseph Mpore is an Agricultural Research Officer at Tanzania Agricultural Research Institute—Makutupora (TARI- Makutupora) in Dodoma. The research institute is in the Central Zone and has been working hand in hand with TARI-Naliendele in groundnut seed multiplication under TL projects. We talked to her in the month of March 2018.

Mpore (Fig. 2.25) says that there has been a lot of changes that the TL projects have brought to her research institute. *"There is an introduction of new varieties within the centre which were not present at the beginning; like Pendo, Mnanje, and mangaka varieties. This has helped our centre to provide new variety to farmers in nearby villages,"* she adds. *"Due to the nature of the environment of the Central zone, groundnut is the second cash crop, making the crop important to farmers here. The availability of the groundnut is hence important for the Central zone."*

Mpore gives us visible differences farmers in the Central zone have identified between the improved and the old varieties. "We can compare in production where the local varieties are not drought resistant, but the new varieties are. They can withstand the dry spell from February and when the rains come they start flowering and the farmer starts harvesting crops as usual but farmers who plant the old varieties end up with zero harvest."

**Fig. 2.25**  Ms. Felista Mpore (Left) with Ms. Nuru James Mgale; both Agricultural Research Officers at TARI-Makutupora's farm, Tanzania (Photo: Ndichu J)

Mpore is satisfied with the uptake of the improved groundnut varieties within the Central zone. *"I can say it is very high that even last year the seeds we had were not enough. If you move across the whole of Dodoma for example, the need for the new variety seeds is very high but the production is still low. Probably it's because there are no farmers that are involved directly in the production of seeds. The seeds are still in the Research Centre,"* she says.

She is calling upon the government of Tanzania to chip in and assist the institute by availing more funds to enhance seed production. *"We know that the production of seeds is key for crop production. If you fail in producing quality seeds you fail to get the quality plant so if the government can support the seed production, I think we are going to fill the gap of seeds we have in the Dodoma region."*

## 2.8   Digitization of Data Collection Practices Becomes the Common Practice Within the Research Programs

Mr. Charles James Mkandawile is a Principal Research Assistant (Fig. 2.26), based in Agricultural Research Institute-Naliendele (ARI-Naliendele). He is currently working as a research technician under groundnut program, and his main activities include trial management. His daily duties involve preparing trials, but sometimes he is engaged in planning those trials, that is, packing seeds and planting. He also does all the data collection, management, and analysis. He has been working on the TL projects since 2007.

Mkandawile is full of gratitude to the TL projects. *"I think I should be thankful to the TL I, II, III projects, during the whole time I have received a number of short-term courses. I attended a course in India, I have also attended several courses in Malawi on disease causes, and I have also attended a course on data management and analysis in Ethiopia and Malawi"* he narrates.

He says that his life as a research technician has significantly changed since the TL projects set in Naliendele. *"Before the start of these projects, we were using conventional methods; using items like calculators, collecting data using plain papers, but TL for instance, has empowered us and introduced items like a tablet. We now have specialized tools to undertake our work effectively. I can say we have*

**Fig. 2.26** Mr. Charles Mkandawile, a research technician, using a mobile application to collecting data on trial at Tanzania Agricultural Research Institute, Makutupora, in Dodoma, Tanzania (Photo: Ndichu J)

*benefited immensely from these projects. Through the trainings and use of improved technologies, I am now able to work more in less time, unlike before."*

Mkandawile says that farmer participation has been of utmost importance. *"I have worked with farmers on on-farm trials, where apart from variety selection; we also use this opportunity to train them on other management practices for improved groundnut varieties. We also do demonstration plots from which we conduct field days, and Nane Nane shows (Agricultural Exhibition) so there is a lot of interaction between us and the farmers."*

## 2.9    TL Projects Make Work Easier for Agricultural Extension Officers in Tanzania

Extension officers across different regions in Tanzania who got involved in TL projects too have success stories to tell. We talked to Ms. Leodina Ernest Mpagama from Mnanje, Nanyumbu District, Mr. Loti Philipo Molleli from Mpeta (Fig. 2.27), Masasi District, both in the Southern Zone, Mr. Joel N. Mpayo from Mindola, Mr. Anthony Sahali from Ilindi, both of which are found in Bahi District within the Central Zone, and Mr. Samson Sumuni from Kahama Mjini, in the Lake Zone. Although from different regions across Tanzania, these five share a common story of how farmers have now been able to improve their livelihoods through embracing improved groundnut varieties.

The extension officers have all gone through training which they transfer to the farmers. For instance, Molleli reported, *"Some of the teaching I remember I went*

**Fig. 2.27** Mr. Loti Philipo Molleli (Left in both photos) explaining disease diagnosis to Mr. Nyirenda, a farmer in Maugura Village in Masasi District, Tanzania (Photo: Ndichu J)

**Fig. 2.28** Ms. Leodina Ernest Mpagama (in dark blue shirt) teaching women and youth farmer groups best practices in groundnut farming at Mnanje, Nanyumbu District, Tanzania (Photo: Ndichu J)

*through from Naliendele is how to help farmers use the improved seeds and planting them using the right spacing."*

In the course of their work, these extension officers meet and interact with many farmers within and without groups. *"I train many groups on how to plant the new improved groundnut varieties, but I deal with more individual farmers than the groups. In Mnanje alone, for instance, I have about 2700 farmers. I train them five days a week especially during the planting season,"* Leodina said (Fig. 2.28). Loti, on the other hand, is working with five groups of an average of 20 farmers each, and over 100 other individual farmers.

In Ilindi, Sahali (Fig. 2.29) says he was able to target special groups, which are now able to meet their needs through the production of groundnut. *"We have a women's group that takes care of orphan children. These women needed a source of income to raise the children, and groundnut came in handy. There is also the youth who are susceptible to moral decay due to lack of funds. Many ladies especially will tend to go to the city and engage in prostitution. To keep them in the village, one*

**Fig. 2.29** Mr. Anthony Sahali (extreme right) with Owe Faraja Group Members in Ilindi, Bahi District, Tanzania (Photo: Ndichu J)

**Fig. 2.30** Mr. Joel N. Mpayo from Mindola (extreme left) talking to Owe Faraja Group members in Bahi District, Tanzania (Photo: Ndichu J)

*need to give them something to do, and what a better way than through groundnut farming"* he says. The role of these extension officers is huge in the passing of knowledge to farmers within the villages and remote communities.

In Mindola, Mpayo (Fig. 2.30) said of a unique approach they have employed. *"Agriculture fraternity within Mindola decided to focus on educating school pupils about these new groundnut varieties because as they grow old, they will have been*

**Fig. 2.31** Mr. Samson
Sumuni going through
records in his office in
Kahama District, Tanzania
(Photo: Ndichu J)

*inculcated with the practice of this kind of farming. They will also be able to pass this knowledge to their parents, and it will be faster to spread this knowledge,"* says Mpayo.

In Kahama, Sumuni (Fig. 2.31) told of the success story of how new businesses to process groundnut in the area have sprouted up. *"After seeing that groundnut are now available in large quantities here, we have business people in the business of processing and grinding groundnut."*

## 2.10   Development Organizations Partnered to Spread Improved Varieties to Communities

American politician and environmentalist, Mr. Albert Arnold Gore Jr., was once quoted saying, *"If you want to go quickly, walk alone but if you want to go far, walk in a team."* The team running the TL projects in Tanzania knows this too well, and they have endeavored to involve like-minded people to help spread the uptake of improved groundnut varieties within Tanzania.

One such group that was involved is an NGO known as Masasi High-Quality Farmers' Products. The NGO is located in Masasi District, and Ms. Samia Noel Seif is the manager (Figs. 2.32 and 2.33). We talked to her in March 2018, and she said, *"We are dealing with cashew nut and other crops and we have twenty-five groups in Masasi District with a total of 6,118 members. We work with TARI-Naliendele and they have helped us since 2008 by providing us the best groundnut seeds."*

She says the improved variety seeds have increased the percentage of groundnut farmers in Masasi District. *"We have three varieties; Mnanje, Pendo, and Johari. Farmers mostly prefer Pendo as it grows well and fast; in just three months."*

Seif reported more about benefits they got under TL projects. *"We received training from TAri-Naliendele on areas like how to space when planting, diseases like aflatoxin that affects groundnut quality and about varieties that tolerate drought. They also buy produce from our farmers and sell them on our behalf, so farmers have a market."*

Seif's parting shot can only be put in her own language in the fear of watering down the meaning, *"Hizi mbegu za kuboreshwa zimeleta manufaa kama kazi ili-yopo shambani imepungua kwa vile wanapanda kwa utaalamu na sasa hawapandi*

**Fig. 2.32** Masasi High-Quality Farmers' Products, Office's establishment in Masasi District, Tanzania (Photo: Ndichu J)

**Fig. 2.33** Ms. Samia Noel Seif working at Masasi High-Quality Farmers' Products, Office in Masasi District, Tanzania (Photo: Ndichu J)

*bora mbegu, wanapanda mbegu bora."* (These improved variety seed s have been beneficial as the farmers work less, grow in expertise and are not growing any seeds but better seeds).

## 2.11   District Authorities Happy with TL Achievements for Their Farming Communities

By now, one may wonder whether agricultural leadership was involved. Yes, it was, in the grassroots. We spoke to three District Agriculture, Irrigation and Cooperative Officers, from different agricultural zones, locally known as DAICO in Tanzania. Mr. Aithan Chaula, from Chimwino District, in Dodoma, within the Central Zone (whom we will meet again later speaking in the capacity of a private company seed producer), Mr. Tamba Wilfred from Masasi District in the Southern Zone (Fig. 2.34),

**Fig. 2.34** Mr. Tamba
Wilfred talking to visitors
at his office in Masasi
District. Tanzania (Photo:
Ndichu J)

**Fig. 2.35** Mr. Aithan Chaula shows us some of the groundnut varieties in his farm in Chimwino District, Dodoma, Tanzania (Photo: Ndichu J)

and Mr. Aron Manguzi from Ushetu District in the Lake Zone. We talked to them between March and April 2018.

They too have all gone through trainings. *"We were trained on innovation, mainly in groundnut and beans and how to assist farmers to change from what they are used to and make them view groundnut as a business,"* says Chaula (Fig. 2.35).

Though from a different region, Tamba shares the same sentiments. *"The benefit from this is that groundnut have been taken as serious business with the new improved varieties. But most importantly, production has improved therefore farmers can meet their basic needs from the groundnut production thanks to TL projects,"* he says.

Apart from groundnut, Tropical Legumes I, II, and III also focused on common beans in Tanzania. In early April 2018, we visited Selian Agricultural Research Institute in Arusha, which is in the Northeast of Tanzania, and about 100 km from the Kenyan border. We also visited Mbeya, a town in the Southwest of Tanzania, just about 100 km from both Malawi and Zambia borders. Here, we found the first collaborator of Michael Kilango at Tanzania Agricultural Research Institute-Uyole who took us to meet different people in Mbeya and Mbozi areas.

## 3.1 TL Projects Enhance the Effectiveness of Breeding of Common Beans in Tanzania

### 3.1.1 Researcher Shares Perceptions of TL Projects' Achievements

Ms. Shida Nestory (Fig. 3.1), an experienced agricultural research officer and common been breeder at Selian, presented the projects' achievements for common bean breeding program in different areas. For accessibility of the new germplasm, new materials can be accessed sufficiently through the International Center for Tropical Agriculture (CIAT), the local collections (landraces), and other national stations (Uyole, Maruku). *"Thanks to TL projects we are now able to advance six generations of bean varieties each year. The new breeds that we are developing have traits like better resistance to diseases, a higher nutritional value (iron), early maturity, and ability to survive drought periods. We have 8 old varieties of beans that are aged more than 10 years since the time they were registered; we also have 7 new registered varieties which gained registration in 2018, also underway is an additional 8 lines that are under multi-location trials. Through TL III, we have been able to release a total of 15 varieties up to date. Of the 15 varieties, five are climbing bean types. The 15 varieties have reached farmers across the Tanzanian farming regions. Currently,*

© The Author(s) 2020
E. Akpo et al., *Sowing Legume Seeds, Reaping Cash*,
https://doi.org/10.1007/978-981-15-0845-5_3

**Fig. 3.1** Ms. Shida Nestory (Squatting left) inspecting crosses in the crossing block at TARI-Selian in Arusha, Tanzania (Photo: Ndichu J)

*we can produce about 7 tons of breeder seeds on an annual basis. TL Project can be said to have played a dominant role in our breeding process, 60% of the resources that we own and use have come from the TL Program. The new varieties are very superior in that they have higher yields; they mature much faster and have a higher nutritional value in comparison with the old ones,"* Shida ended.

### 3.1.2    Nutrient-Dense Common Beans Available to Improve Malnutrition

The varieties that we develop have different quality traits, but we have noted that there is a great affinity of bean seeds that are rich in iron and zinc. This is because malnutrition is very high in this region, and thus the need for beans that improve the nutrient quality of meals. There is also preference of bean varieties that can withstand harsh weather conditions such as drought without leading to huge losses in the farming process.

### 3.1.3 TL Projects Have Increased Research on Beans Seed Systems

The TL projects have played a significant role in the seed sector improvement scientifically. Prior to the TL projects, the research conducted on the bean seed was very little and bore very minimal outcomes on improving the quality of the bean seed. Ms. Edith Kadenge (Fig. 3.2) who is a seed researcher and the coordinator of the TL projects at Selian Research Institute in Arusha Tanzania does not shy away from stating the positive changes that have emanated from the TL projects. Following the inception of the TL programs, there has been a great increase of seed demonstrations conducted in the regions that grow bean seeds. Kadenge reported to us during our study that there has been an average of 15 demonstrations conducted in every district that grows beans meaning that there has been a great outreach to farmers in the regions around Arusha, Tanzania.

Ms. Sylvia Monica (Fig. 3.3) who works with CIAT-Arusha stated that they have been able to release over seven varieties of beans which have improved traits to the farmers. According to Ms. Monica, under TL projects, the institute released varieties of beans once in every 2 years.

Before the TL projects, she says, there were no seed companies that were involved in the production of the bean seed, but after the rollout of this project, many companies have taken up the role of production and sale of seeds. In collaboration with us, companies that solely produced maize seeds have taken up the beans seed production role. The Tanzanian government has also taken up the role of improving quality and quantity of bean varieties in Tanzania and through the Agricultural Seed Agency across the country.

According to Monica, ever since the TL projects started in 2007, the production of beans seed has improved immensely. In 2015 alone, for instance, there was a production of over 104.2 tons of beans seed produced by different institutions within Tanzania. Ms. Monica says this is a very high number as compared to past years. Noting that there was no single seed company that was involved in the production of bean seeds around Arusha, the current five companies and over 80 groups of farmers that are solely involved in bean seed multiplication and sale is a great milestone. This development can be attributed to the TL projects that brought the much-needed changes.

**Fig. 3.2** Ms. Edith Kadenge, Researcher, Selian Research Institute in Arusha, Tanzania, shares how many seed companies have taken up the role of production and sale of improved bean variety seeds (Photo: Ndichu J)

**Fig. 3.3** Ms. Sylvia Monica, MLE officer, CIAT-Arusha, Tanzania, shares her insights in bean production (Photo: Ndichu J)

Through our study, we realized that women involvement in beans farming is high. Ms. Monica told us that beans were known to be a crop for women in Tanzania. Despite the huge improvement and commercialization of the seed, in recent times women are the ones who mainly deal with the farming process. It was however noted that males take up the produce after harvest and do the selling. Ms. Monica reported that they had been able to reach to over 5000 female bean farmers in the southern and northern zones in Tanzania, therefore noting that the number of women farming beans is higher as compared to the males involved in the same.

In the past, farmers were not able to access improved beans seed since there were no proper networks for this to happen. The government body tasked with production and distribution of bean seeds in Tanzania is the Agricultural Seed Agencies (ASA) which was not in a capacity to supply the demands of the farmers in Tanzania. Through the TL projects, farmers are now able to access the seeds that have been improved, and in turn the overall farming of the bean seed has improved greatly, and this is said to go on in the future leading to the improvement of the lives of bean farmers in Tanzania.

The spread of new data on bean growing is more accessible to the farmers, as compared to days before the inception of the TL projects. Kadenge states that there is the use of local radios; for instance, farmers in the Northern Zone in Tanzania get access to information on beans through Radio: Sauti ya Injili, Radio Utume Fm, and Habari Njema FM that broadcast across the region. This use of the radio broadcast system has been aided by organizations such as the Farm Radio International which aims at disseminating useful farming information to farmers.

Through radio shows, researchers and agricultural stakeholders under the TL projects have successfully offered information on the best farming practices from the initial planting process to the final stages of production, and on storage of beans after. As Kadenge reports, the outcomes of these programs have been great, and farmers are harvesting better crops and hence more profits and benefits from the beans they plant. On average there is a direct reach of over 50,000 farmers annually in the Northern Zone through the media. This number is great as it causes a ripple effect on the state of bean farming in the entire region. In recent time there has also been an increase in dissemination of better farming practices of beans using local television broadcast in Tanzania. The overall effect of the use of media in promoting better farming of beans can be seen everywhere; Kadenge emphasized the fact that all these efforts have emanated from the TL projects.

There are many training sessions that the team led by Kadenge have conducted to farmers across the Northern Zone about beans. The farmers have been taught on the best ways of multiplying the new breed varieties that are provided to them by breeders. There are many new bean variety seeds that have better traits which have been recommended to farmers. For instance, the information has helped farmers in Mburu, Babati, Arumeru, Siha, Hahi, Same, and Moshi regions of Tanzania to enhance agronomic practices. These demonstrations have enabled the creation of networks with seed producers, agro-dealers, and NGO's that have at the end brought benefits to the farmers. *"We have been supporting production of Mark 44 and ARA W3225 (they are climbers and require support). Farmers are happy with CAT P1 and Ngolole varieties which they prefer and say that it cooks fast and is palatable,"* Ms. Kadenge reported.

In light of the endeavors to improve bean farming in Tanzania, there are numerous challenges that people in this program have faced. Ms. Kadenge stated that one challenge in improving the breeds of beans is that the farmers who at times have low economic ability tend to sell all the seed that they are given for multiplication due to poverty leading to a discontinuity in the growth of the new breeds. Climatic challenges also affect the promotion of better bean planting, for instance, drought at times leads to huge losses to farmers. As Kadenge reported, the demand for the improved bean seeds at times is higher than the capacity to produce leading to shortages and thus farmers resort to planting the old varieties. A few seed companies have been skeptical in the production of new varieties which has hampered the speed of seed multiplication.

### 3.1.4 TL Projects in Tanzania Factor in Issues of Gender in Bean Production

As previously seen in this report, the bean seed was previously considered a crop for women and one that was planted for subsistence use only. Following the inception of the TL projects in Tanzania, a lot has changed on the issue of beans and the gender aspects that are related to it. To understand this comprehensively we sought information on changes that are related to gender that have emanated from the TL projects involvement in bean seed systems. We interviewed Mrs. Ms. Eunice Zakayo (Fig. 3.4), a gender and social economic expert from Tanzania. Zakayo has studied the issue of gender and its connection with legumes in Tanzania.

> She stated, *Based on the value chain, we have different stages of the seed production and in those stages, there are different opportunities. Men are mostly owners of the farms, so they are the decision makers. They prepare the fields with tractors or cow ploughs, spray herbicides for weed control, and ferry materials from the farm to the stores. They are also involved in pricing. Women, on the other hand, do not have much say in these matters. Actually, women just intercrop for home food and seed production. During planting, they use ploughs while men sometimes broadcast the seeds. The women also weed the crops, twice or thrice depending on the field condition. They often do this in groups. Harvesting*

**Fig. 3.4** Ms. Eunice
Zakayo, Gender scientist,
Selian Agricultural
Research Institute, Arusha,
Tanzania, shares about
involvement of women in
revolutionizing bean value
chain (Photo: Ndichu J)

*the crops and piling is also done by women as they wait for the men to ferry the harvest.*
*They make sure the harvest is dried and packaged and wait for the decision maker to come*
*with a buyer.*

With Zakayo's response, it was clear that the gender involvement in beans production had changed, thanks to the involvement of TL project in beans. The beans seed as Zakayo told us in the study was mainly produced in small scale and was just farmed for home use. In the past, men did not involve themselves at all in the farming of beans. This has however changed and now there is a lot of division of labor in beans production as business where men are now taking up the role of major decision-making such as determining the amount to be planted and also negotiating the sale of the produce. The men also get involved in the masculine tasks that are involved in the bean farming process. The women, however, do the better part of the work in farming of the bean crop.

In the study, Zakayo told us that, *"Sometimes men prefer the marketable crops while women prefer high yields and palatable crops. Decision making has improved especially in Kagera. They get their training in groups, share the farms, learn together, and adopt the teachings into their community which turn to high yields. They know which fertilizers to use and what to do in which stages of crop developments. Ways in which they spend the money have seen these women's households improve and get access to better seeds. Women have better results from what they do on the farms. Formerly, women did not have any knowledge on farm inputs on certain varieties or did not have access to information. Now they can get the information, network while looking for a market, and have groups which help them organize themselves and assess what they do."*

The betterment of the bean seed systems has emanated from the efforts of the TL projects. There has been a transition of bean farming as a subsistence crop to a commercial crop, and this has in turn led to the economic empowerment of the women. Before, the women were not able to make as much money that they make now from the seed farming. This fact indicates that female farmers have now become financially stronger like their male counterparts. Zakayo reported to us that men are now appreciating the work done by the women and have since started working hand in hand with the women. The overall ripple effect is that females have gained more appreciation in the society.

Zakayo brought about the issue of resistance to adoption of new technologies in the bean farming process due to cultural norms. She stated, *"There are cultural barriers which hinder us from getting the most out of these innovations. We have people on the ground that are helping us tackle these issues in driving women forward and everyone has a role to play in making sure that no one's beliefs or norms have been violated in the process. We promote collective decision making among a man and a woman which makes a strong household. We have multi-stakeholders who involve different groups of farmers, who are invited to see machines and help them understand the markets and approaches."*

In view of Zakayo feedback, women at times are hindered from making so much advancement economically, adoption of farm mechanization being one of those things that seem to go against the norms in the society in Tanzania. Through the TL projects, there are plans to suppress these norms in a way that demean the progress of females in Tanzania. When we asked about the efforts that were in place to empower or change the situation of women who farm beans through the TL projects, Zakayo responded that, *"We have multi-stakeholders which involve different groups of farmers who are invited to see machines and help them understand the markets and approaches. We also found that women are aggressive in asking questions. We also found that women numbers are growing so when we go to the fields, we encourage them to go to markets and get information on prices and compare them. They are now motivated and aggressive in getting information. We have platforms for marketing and seed production. This makes them more aware of what is going on. For instance, there is a lady who has been growing beans for several years, and before she got the training, she used to get very little harvest. She is now doing really well as the produce has increased. We often go there to promote her and use her case for demonstrations. She now supplies grains to a wildlife reserve despite not having the capital."*

Following the rollout of the TL projects, there is a notable paradigm shift in the relation and value of genders. Prior to the projects the females were seen as less productive, but after the positive changes that emanated from these projects, women have become more financially able and are improving the lives of their families in ways that they could not do before. With this, the men are seen to respect the women more and even involve them more in major family decisions unlike in the past.

There are some challenges that Zakayo reported which were an impediment to the good progress incepted by the TL projects. The quantity of seed that the farmers require is still not adequate and the multiplication process is still not efficient.

## 3.2    Research Technicians in Common Beans Production Get Specialized Tools of Work and Trainings Under TL Projects

The process of ensuring that the best varieties with high yield, better resistance to drought, pest and diseases, as well as high nutritional value requires so much inputs to achieve. To understand the technicalities involved in the bean seed production in Tanzania, we interviewed Mr. Alex Christopher Kisamo who is a research technician. Kisamo has been in this field for the past 15 years. He was quick to point out that a lot has changed ever since the TL projects were incepted in this region. He stated, *"I have been involved in the TL Projects ever since they were incepted in this region. I was taken through training during the project that took place in Ethiopia, where I learned how to select the best beans varieties and guide the farmer on the best seed for planting. Later I was taken through another training in Malawi where I gained more experience in seed selection process. I can say that prior to this project, it was very difficult for people like us to gain the necessary skills that we utilize while conducting research and training farmers."* The people who are responsible for training the farmers and conducting research require frequent training, and as Kisamo revealed, this only happened after the TL projects were incepted.

Kisamo also noted that in past years, there was limited access to the necessary equipment in research in the beans field. He told us, *"The equipment that we lacked for conducting our research process was availed to us after the inception of the TL Projects. We were able to obtain quality weighing machines, soil analysis machines and other equipment that we use in the research process. I can attest that through the acquisition of this equipment, our work has become easier and the results are better. Initially, we would take a lot of time to develop a bean variety; however, after the availing of the new equipment we have been able to develop eight new improved varieties in a very short period."*

There has been seven new bean varieties that have been developed since the inception of the TL projects in 2007. Kisamo (Figs. 3.5 and 3.6) stated that the equipment they had in the past were outdated and gave them an uphill task in the research process. The equipment that they had prior to the TL projects did not

**Fig. 3.5** Mr. Alex Christopher Kisamo using a digital device to collect data on beans in Arusha, Tanzania (Photo: Ndichu J)

**Fig. 3.6** Mr. Alex
Christopher Kisamo
displays a tablet that TARI
Selian in Arusha, Tanzania,
acquired through TL
projects (Photo: Ndichu J)

provide instant data and took them days to produce results. He noted that thanks to the TL projects, they could take samples from the ground, run tests, and get accurate and instant results. The trips to and from the lab to synthesize data have since been done away with. Kisamo stated that the logistic costs had also become lower, and in turn the cost of research went down greatly as the results went up. The effect of this better equipment and training of stakeholders in the research process has resulted in the increase in quality of the beans being produced. The beans being harvested by farmers have better yields and are of more nutritional value. Kisamo states that if it were not for the TL projects, the quality of bean being planted and consumed by farmers would still be very low. He did not shy away from expressing his appreciation of these projects and insisted that the legume sector could be improved to the optimum level leading to overall growth of the Tanzanian economy that depends heavily on farming.

Challenges are still there as Kisamo told us. He stated, *"Our production of the improved bean seeds is still very low in comparison with the demand by farmers. Due to the evident value of the beans and its increase in price at the market, there has been a great demand by farmers who seek to reap profits from the new varieties; this is a challenge to us in the multiplication of the seeds of varieties that we develop."*

## 3.3   Research Centre Directors Attest the Increase of Resources and Infrastructure Under TL Projects

Following the inception of the TL projects, there has been many changes seen in the research centers. Those that run these institutions are not shy of expressing their joy for to the positive changes brought by TL projects. One such person is Rama Ngatoluwa, research coordinator for the Selian Agricultural center. He spoke to us on behalf of the institute's director. Selian is the headquarters for the Northern Zone research center. It has a sister center that deals with horticultural research. The center has three research departments: one that handles soil fertility issues; another one that deals with all crop-related research such as breeding, weed science, and post-harvest; and the department that links research and extension for packaging information and taking it to the intended audience.

We were also able to get an exclusive interview with one of the research center directors, Dr. Tulole Legendo (Fig. 3.7). Dr. Legendo, the zonal director of research and development in the Southern Highlands, based in Tanzania Agricultural Research Institute (TARI)-Uyole, within Mbeya, Tanzania. He mentioned that due to the expansiveness of his zone, it covers nearly all climatic conditions and thus all plants that can grow within East Africa grow in this zone. Dr. Lugendo pointed out that TL projects had played a big role in improving farming of beans in Tanzania. *"TL Projects has facilitated quick research that has enabled TARI-Uyole, for instance, to develop new bean varieties,".* *"The research center has released over 23 types of beans varieties that have better traits than their predecessors,"* he added.

Like Dr. Legendo, Dr. Ngatoluwa (Fig. 3.8) was also very eager to express the benefits that emanated from the TL projects. He stated, *"Our institute started engaging in Tropical Legume projects in 2007 where we did the projects in phases. Our engagement with the projects has enabled increases in productivity. We have broadened the variety of seeds in the country and have also been able to increase the nutrition value of the product; the bean has seen an increase in zinc and iron content."*

On infrastructural advancement, Ngatoluwa said, *"TL projects helped us improve on transport and stationary equipment"* (Fig. 3.9). The TL projects are seen to have brought many changes in the phase of agriculture in Tanzania as Dr. Ngataluwa rightly say.

Despite the huge success of the project, there are also a few challenges that still face the bean industry as Dr. Ngataluwa emphasized. *"The major challenge is the increase in demand that at times we are unable to meet. Also, seed accessibility by farmers is an issue as we are not able to reach all the farmers due to lack of proper infrastructure like adequate roads. Sometimes the production of the seed is very low compared to the outputs due to certain issues like diseases and pests and weather conditions."*

This notwithstanding, like most stakeholders in the beans sector, Dr. Ngataluwa closing remarks were credits to the TL projects and stated that since its inception, the farmers have been able to meet the basic family need that they could not meet before due to the production of the improved varieties.

**Fig. 3.7**  Dr. Tulole Legendo, Zonal Director of research and development, Southern Highlands, TARI-Uyole, Mbeya, Tanzania (Photo: Ndichu J)

**Fig. 3.8** Dr. Rama Ngatoluwa, the zonal director of research and development, based in TAnzania Agricultural Research Institute, (TARI)-Selian, Arusha, Tanzania (Photo: Ndichu J)

**Fig. 3.9** Printers and laptops, some of the other equipment Selian Institute in Arusha have benefited from TL III project (Photo: Ndichu J)

## 3.4   Agricultural Extension Officers in the Bean Crop Tell of Their Benefits Under TL Projects in Tanzania

Extension officers in Tanzania have also benefited a lot from the TL projects. Mr. Clay Salehe Sarumbo (Fig. 3.10) is an agricultural officer in Ruanda region, Mbozi District in Songwe Province. Sarumbo narrated the progress before and during the implementation of TL III projects in bean farming in this region. He stated, *"I remember when we started the project in 2007, we were able to sell only 5 tons of bean seeds at the time and a kilo of the seed was going for 1050 Tanzanian shillings. The following year the amount increased greatly to over 28 tons and 33.5 tons the following year. Following the introduction of the improved varieties, our farmers have been able to reap between 600 to 900 kilos per acre where in past years they would only get 100 kilos."* Sarumbo noted that following the inception of the TL Program, the productivity of the bean seed has increased tremendously, and farmers are making fortunes from their farm which was not the case in the past where beans were just a subsistence crop.

In his region alone Sarumbo reported that he is involved in the training of over 171 groups of farmers, and the major form of training that is given by Sarumbo is through field demonstrations. He stated that the farmers have been very cooperative upon seeing the good returns from the improved variety seeds. The farmers are

**Fig. 3.10** Mr. Clay Salehe
Sarumbo, an agricultural
officer at Mr. Daudi
Bukuku's Homstead in
Ruanda, Mbozi District,
Tanzania (Photo: Ndichu J)

**Fig. 3.11** Mr. Clay Sarumbo (extreme right) at Mr. Daudi Bukuku's home in Ruanda, Mbozi District, where sun drying of beans was taking place in early April 2018 (Photo: Ndichu J)

quick to inquire in case they feel they need professional assistance from the likes of Sarumbo. This situation had never been witnessed before and Sarumbo credits this development to the TL projects.

However, Sarumbo (Fig. 3.11) has observed as an extension officer that more still needs to be done in changing the farming culture in Tanzania. Changing people's mindset is always difficult and requires a lot of persistence. Since farmers were deep into the past ways of farming beans, making them apply the new farming techniques and seeds is at times met by rebellion. Also, due to the good performance of the improved varieties it has been difficult for the seed producers to meet the demand in the market. Sarumbo hopes that all the relevant bodies including the government will aid in production and distribution of the improved beans and thus alleviate the shortages seen now.

## 3.5   Non-Governmental Organizations Working Hand in Hand with TL Projects to Develop Production of Common Beans in Tanzania

There are many NGO's that are involved in the improvement of agriculture in Tanzania. In our study, we interviewed Ms. Jacqueline Sanga (Figs. 3.12 and 3.13), a Monitoring and Evaluation Officer at *Action for Development Program* (ADP), an NGO in Mbozi, Tanzania, that was started in 1986. ADP is involved in educating farmers on the best farming practices that are there and how they can conduct farming in a sustainable manner. Sanga stated, *"We were initially focused on the production and storage processes in farming improvement but when we realized that farmers were now able to successfully revamp these stages in farming, we moved to the marketing sector. At the end of the day, we seek to improve farming in order to bring profits to the farmers in this region."*

Sanga noted that her organization was working with farmers from over eight regions in Tanzania. She told us that the TL projects have been very instrumental in their endeavors to improve agriculture. Sanga stated, *"we have worked with TARI-Uyole in many instances to aid farmers in improving their farming. We were able to access bean seeds that had been heavily researched and invested on by TARI-Uyole. We promoted these varieties to the farmers and this led to the huge success in bean farming which is the current situation."* She emphasized how the TL projects were key to improving the state of bean farming and the whole agriculture in general. Since the TL projects started, Sanga stated that the organization had benefited a lot, their extension officers got training which they, in turn, disseminated to the farmers. She noted that as an NGO they were not able to develop improved bean varieties and thus it was a great contribution by TARI-Uyole through the TL projects that allowed research and development of beans.

*There has been a creation of a network such that the TARI-Uyole team does all the development of varieties and us and other NGO's channel the seed to farmers,* she told us.

Sanga added, *there are challenges that we have faced in the course of our work; climatic conditions sometimes are not in favour of the bean seed farming and there are also issues of crop diseases and pests' infestation leading to losses. Therefore, more research is welcome.*

**Fig. 3.12**  Ms. Jacqueline Sanga, Monitoring and Evaluation Officer at Action for Development Program (ADP), an NGO in Mbozi, Tanzania (Photo: Ndichu J)

**Fig. 3.13**   Establishment of ADP Mbozi offices in Mbozi District, Tanzania (Photo: Ndichu J)

## 3.6    Private and Public Seed Enterprises Flourishing in Common Bean

### 3.6.1    Private Seed Companies Venture into Multiplication and Distribution of Improved Bean Varieties

Like is the case with groundnut, the seed business has attracted major players in the industry. Private individuals have set up businesses to fill the gap left in the market. As the popularity of beans grew, demand grew as well and since demand had not been adequately met, a business opportunity cropped up and the private sector took advantage of this. To give more insights into this, we interviewed Abel Samuel Byda, the Managing Director of Byda Agrovet Co. in Mbulu, Manyara District (Figs. 3.14 and 3.15). The company has five different outlets in Tanzania and has employed 24 workers. Byda specializes in selling seeds directly to farmers. He started this business 5 years ago and has been going strong ever since. He gets his seeds from TARI-Selian. He also has a 5 ha piece of land that he has planted to beans.

*"My living situation all round has drastically improved since I am getting good profits. I started off with one retail shop, but I have now expanded to four more shops. My harvests are growing gradually as seasons go by."* He told us that his clientele is exclusively farmers since they are the ones who use his produce largely.

We also met Ms. Cecilia Magesa, Meru Agro's Regional Manager in Arusha. She stated that the company produces various types of beans varieties which are distributed in the north part of Tanzania. They have been largely involved in the production of two varieties, namely the *Yamugo 90* and the *Uyole Njano*. Meru Agro (Figs. 3.16 and 3.17) has been in business for 7 years where together with TARI-Selian they have worked hand in hand in getting parent seeds and sponsoring activities such as breeding the varieties and looking for seed markets. *"We have also gotten help from NGO's which have bought these seeds from our region and taken them to Kadira (this region does not get much of the seeds). Our work in the south is not much since the weather over there is not favorable,"* she says. In the last harvesting season, they were able to produce 120–200 tons.

**Fig. 3.14**  Establishment shot of one of the Bayda Agro-Vet co. Shops in Mbulu, Manyara District, Tanzania (Photo: Ndichu J)

**Fig. 3.15**  Mr. Abel Samuel at one of his shops in Mbulu Manyara District, Tanzania (Photo: Ndichu J)

This, however, is not reflective of the challenges they face since it is way low from the set goals. The crops are often attacked by diseases and ravaged by the weather. *"Our customer base is really vast since we sell our products to institutions and NGOs,"* she concludes.

**Fig. 3.16** Establishment of Meru Agro-Tours & Consultants Co. Ltd. shop in Arusha Mjini, Tanzania (Photo: Ndichu J)

**Fig. 3.17** Ms. Cecilia Magesa, Meru Agro's Regional Manager in Arusha at the Company Shop, Tanzania (Photo: Ndichu J)

### 3.6.2   Public Seed Companies in Tanzania, Agricultural Seed Agency (ASA) Takes Up Multiplication of Improved Varieties of Beans in Partnership with TL Projects

As mentioned earlier, the government of Tanzania, through the Agricultural Seed Agency (ASA), has been able to catalyze the efforts by TL projects in multiplying and distributing new improved bean varieties in the country. This public institution is run by agricultural officers. Their duties are multiplying seeds for the purpose of satisfying the seed demand around Tanzania. We managed to have a sit down with Mr. Eliud J. Musumi (Fig. 3.18), an agricultural officer (ASA-Mbozi Farm) who emphasized that their main goal is seed production. *"Previously we only dealt with maize seeds, but we decided to expand our horizons into bean seed production,"* added Musumi. *"Our decision to include bean was a success."*

Last season, the agency planted 45 ha of land but they decreased this number to 10 ha due to high amount of labor required. After harvesting the crop, Mr. Musumi explains that the product undergoes certain stages, the first one being preparing it for storage. The beans are taken out of their pods and stored in large stores. *"We finally weigh them where some weight is lost due to the drying process that takes place."*

We also visited ASA-Arusha where we met up with Mr. Marco Martin (Figs. 3.19–3.21), a breeder at the government-owned agency. Here, they plant a variety of beans and their produce has seen a gradual increase over the period that he has been working. Martin expects close to 200 tons of beans this harvesting season. He told us that they have planted on over 500 acres of land in which they have designated some part for maize and the other for beans. *"We collaborate with research institutes who advise us which variety works best on certain conditions in order to optimize our harvests,"* he says.

Marco adds that they try to make their seeds accessible, so they look into their distribution channel down to the ward level. They even collaborate with farmers to try and sell the seeds. Last season they farmed on 30 acres with the farmers who received subsidies. When it comes to selling their produce, they package the seeds into 2 kg to 50 kg packages which they sell to farmers.

**Fig. 3.18**  ASA Mbozi Offices in Mbozi, Tanzania (Photo: Ndichu J)

**Fig. 3.19** Mr. Marco Martin, a breeder, ASA-Arusha, at the institution's farm, Tanzania (Photo: Ndichu J)

**Fig. 3.20** Workers at ASA-Arusha in improved Bean farm at the Tanzanian Government-owned institution (Photo: Ndichu J)

Marco is candid in sharing what he goes through in form of challenges. The seed demand curve often dips without any indication. Farmers can go through losses due to this uncertainty. Their farms are surrounded by pastoralists who more often than not lead animals onto their lands for grazing causing conflict and losses. These are some of the encounters they go through, but all in all great millage has been covered through their efforts as an agency.

**Fig. 3.21**    Establishment of ASA-Arusha Farm in Arusha, Tanzania (Photo: Ndichu J)

### 3.6.3    Agro-Dealers Find Business in Improved Bean Varieties Through Efforts by TL Projects in Tanzania

The sales and supply business cannot be completed without the retailers or, in this case, agro-dealers. To get a firmer grip on this, we talked to Ms. Frazia Mbwaga (Figs. 3.22 and 3.23) in Mbeya town. She specifically sells bean seeds in a shop that is part of the chain of shops by Beula Seed Company within Tanzania. Mbwaga who handles two types of seeds *Uyole Njano* and *Uyole 96,* packages her produce as well. She started doing this mid last year and has gone on to become a revered retailer. The relationship that Mbwaga has formed with her customers has made business so easy that she can now operate on credit. She goes on to inform us that beans are a seasonal crop, so she tries to get the most out of it when the time comes. *"The market drifts for this product and you need adequate preparation far the credit to come in handy,"* she added.

Demand also varies; when the market is saturated with beans, demand dips, making retailers almost go out of business, but when the beans are scarce, the demand is high which also has its negative effects. Despite the trying times which can happen for any business, Ms. Mbwaga has managed to stay afloat for the most part. She packs her product in 2 kg bags to 50 kg bags. Her clientele ranges from retailers to farmers.

**Fig. 3.22** Beula Seed Company Agrovet Shop in Uyole, Mbeya, Tanzania (Photo: Ndichu J)

**Fig. 3.23** Ms. Frazia
Mbwaga at Beula Seed
Company Agrovet Shop in
Uyole, Mbeya, Tanzania
(Photo: Ndichu J)

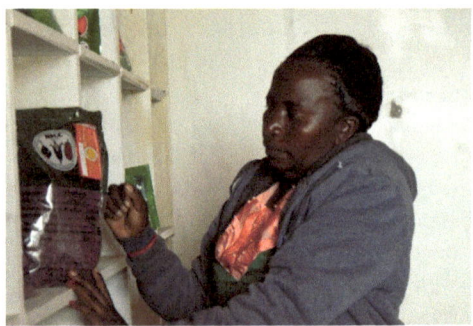

## 3.7    Farmer Groups Improve Their Livelihoods Through Production of Improved Bean Varieties in Tanzania

Groups offer help to young farmers who do not have as much high level of reach as that of established farmers. This often encourages more people to delve into the agricultural sector. We caught up with Zinduka Women Group in Mshewe, Mbeya. Formed in 2011, the group has 20 members, all women, and is chaired by Ms. Witness Sikayanga (Figs. 3.24–3.26).

She says, *"Apart from growing beans as a group, we also grow individually though we put a target of about 600 kg per acre per person."* The women were part of individuals who benefited from the education provided on farming techniques back in the year 2013. They produce seeds with the sole purpose of selling them. They normally plant seeds according to their demand in their local market. They were once involved with the traditional varieties of bean but decided to jump ship

**Fig. 3.24**  Zinduka
Women Group chairperson,
Ms. Witness Sikayanga
shows us a drier at Mshewe
Village Aggregation Center
(VAC) donated by Rafael
Group Ltd and other
partners in Mshewe,
Mbeya, Tanzania (Photo:
Ndichu J)

**Fig. 3.25**  Electricity connection at Zinduka Women Group chairperson, Ms. Witness Sikayanga's
homestead in Mshewe, Mbeya, Tanzania (Photo: Ndichu J)

since they did not have good yields. The newly released varieties have had a positive
impact on the farmers. Since the group solely involve women, changes around them
have been visible to the general public. They can now cash in on home amenities
such as improving the level of their living standards, buying more arable land for
farming, and connecting their homes with electricity (Fig. 3.25).

The road has been bumpy though; the bean market has been monopolized by the
sole buyer who ends up setting up poor prices on beans and the technology to help
in the growth of beans is also expensive for the group.

**Fig. 3.26** Ms. Rhoida Nsagaje, a member of Zinduka Women Group outside her house which she built through proceeds from Beans farming, Tanzania (Photo: Ndichu J)

## 3.8    Bean Grain Market Fed with New Varieties

### 3.8.1   Grain Farmers Embrace New Bean Varieties After Rollout of TL Projects in Tanzania

Small seed dealers often come up in new business environment. They normally limit themselves to a small-scale supply of grains. Such is the case for Ms. Neema Dick Malasusa in Nanyala, Mbozi (Fig. 3.27), who started farming in 2008. She reported that she initially started by planting the traditional varieties which ended up having a detrimental effect on her income. She then switched to the improved varieties which have seen her revenue stream take a turn for the better.

**Fig. 3.27** Ms. Neema Dick Malasusa selling beans in Nanyala, Mbozi, Tanzania (Photo: Ndichu J)

### 3.8.2   Processors in a New Niche in Improved Bean Varieties

Achievements made in improving bean varieties in Tanzania cannot be overemphasized. With such advancements come new opportunities like value addition. This is where processors come in. Ms. Andusamile Mbandile (Fig. 3.28), secretary of Zinduka Women Group, which we featured here, reported that they too have ventured into value addition of the bean. *"We mostly handle the new varieties which we first pluck from the plant, get them out of the pods, clean them and boil them in salty water,"* says Ms. Mbandile. *"This takes place for about 30 minutes then we take them out to dry in a bid to improve their quality in preservation. This type of quality has its pull-on consumers since we improve its taste."* The women have been able to increase their income through this venture, and they say, there is no turning back.

### 3.8.3   Beans Lovers in Tanzania Now Consume Improved Varieties Courtesy of TL Projects in the Country

Beans have garnered popularity among the folks of Tanzania. There has been this notion by the elderly in Tanzania that a meal is not complete if it is not accompanied with beans. Most households in Tanzania often have the crop in their daily meals, a practice our team also witnessed in most restaurants. We talked to Ms. Christina Danson Kalupale (Fig. 3.29), a telephone operator at ARI-Uyole since 1978. Kalupale secured a tender to run catering services in the institution about 6 years ago.

She informs us that through her extensive experience in the catering industry, beans have more consumers than meat. Christina sells cooked food and she reported that the new improved varieties attract more consumers than the old varieties. She has the variety locally referred to as *Kablanketi*, which is preferred because of its thick soup and aroma. This has had positive effects on her business because she

**Fig. 3.28**  Ms. Andusamile Mbandile (right), secretary of Zinduka Women Group, with other group members at their farm in Mshewe, Mbeya, Tanzania. (Left) Members of Zinduka Women Group at their Farm in Mshewe, Mbeya (Photo: Ndichu J)

**Fig. 3.29** Ms. Christina
Danson Kalupale at the
Kitchen in her cafeteria at
ARI-Uyole in Mbeya,
Tanzania (Photo: Ndichu J)

grows the beans she sells, hence optimizing her income. She goes on to state that
during her entire time in the business, she has been handling the new varieties.
Christina adds that the yellow variety is specially requested because it gets ready
quickly and has a sweet taste. These beans have played a huge role in her life as far
as providing for her family's basic and development needs are concerned.

## 4.1 Women's Battle for Financial Independence: How Women Are Using Groundnut to Attain Financial Stability in Uganda

The role of women in traditional African homes is often considered subordinate than that of men. Women in most cases are expected to look after the household and the children and ensure food security while men, on the other hand, are tasked with ensuring financial security. In Uganda, women contribute to 53% agricultural labor force; this is because they have limited access to land and thus resort to offering labor to farm owners. Tropical Legumes projects has strived to empower women through creating awareness and sensitizations to women groups in various parts of the country (Fig. 4.1).

Purlonyo Women Group is among farmer groups that have benefited from the TL projects. Rather than staying at home and waiting to be financially supported, the women have settled on self-empowerment. Ms. Leonora Okidi founded Purlonyo Women Group with an aim of inspiring women to uplift themselves.

The group consists of 35 groundnut producers from Pader District, nine are men despite being a women group. Before joining the Tropical Legumes projects the group produced old groundnut varieties solely for household consumption.

*We started out as farm laborers where we would get hired by community members to till land during planting and weeding season. We also planted groundnut for grain but our earnings from the work were insufficient and we needed to look for diverse ways to sustain our needs.*

Encouraged by the extension officers and project staff, the group members embarked on a journey of farming improved groundnut varieties such as Serenut, 5, 8, and 9 which they received through Green Globe, an NGO that works closely with the Tropical Legumes projects. Ms. Leonora Okidi, the group's chairperson, volunteered 10 acres of her land to the group members for groundnut production. Since

© The Author(s) 2020
E. Akpo et al., *Sowing Legume Seeds, Reaping Cash*,
https://doi.org/10.1007/978-981-15-0845-5_4

**Fig. 4.1** Ms. Apiyo Hellen, a groundnut trader, shows off her roasted groundnut at Arapai market in Soroti, Uganda (Photo: Manyasa E)

joining the TL projects 2 years ago, Purlonyo women group has benefited immensely, besides the improved variety seed, they have been trained thoroughly on production and have been linked to markets for their produce.

The differences between the old varieties and the improved varieties of groundnut are immense according to Leonora. The old varieties, despite having a higher oil content which makes them more desirable, are less resistant to drought conditions, pests, and diseases and are low yielding. The market demand for the improved varieties, moreover, is higher than that of the old varieties. In 2016, for instance, a sack of unshelled groundnut of improved groundnut was 120,000UGX but in 2017 the price went higher to 140,000UGX (1US$ is about 3700UGX at the time of data collection).

> When we started off, we were a bit skeptical on who would buy our produce. Two years into the business, we have links with seed companies and organizations like the Lutheran World Foundation who have promised to buy seed from us after the harvesting season as we produce quality seed. We have also benefited from workshops and demonstrations organized by the TL projects and thus expanded our networks, the opportunities we have received surpassed our expectations.

Since joining the project 3 years ago, the group members' lives have been transformed significantly. Most of the members can now afford to keep their children in school with the profit made from the groundnut sales. Leonora for instance no longer relies on her estranged husband for any form of financial support as she makes enough to send her children to school and cater for all their needs.

Ms. Ademun Loy is also among the many women that have benefited from Tropical Legumes projects in Uganda. Loy, a farmer and small-scale businesswoman, produces groundnut paste and roasted groundnut for sale. Her journey to financial independence started in 1998, when she moved to her sisters' home in Teso Inn village, Soroti District, after she parted ways with her husband. Being the sole breadwinner, her income from her cassava business could not sustain the needs of her two children and herself. Her sister was a groundnut dealer; she roasted groundnut and made groundnut paste which she hawked in Soroti town and sell to neighbors in her home village. *"I mentioned to my sister that I needed an extra source of income since my estranged husband wasn't helping with raising the children, I also needed to get a place of my own as my sister's house was small for both our families,"* she added.

Loy's sister trained her on proper preparation of the groundnut for sale, and in 2000, with a capital of 50,000UGX, she started her own business. She utilized *Erudu*, a white local variety, to make the roasted groundnut and Serenut 2 for making groundnut paste. When she started off, she sold the groundnut to neighbors and hawked some around town, but she currently supplies supermarkets and local shops in Soroti with her products. She makes approximately 44,000UGX a day from 6 kg of groundnut.

In 2016, she acquired home-saved seed from farmers in her village and she planted it on her one acre of land. That season, the crop was adversely affected by Rosette diseases and she only managed to harvest 126 kg of unshelled groundnut. *"The variety I used is an old variety and it is not disease resilient and that is why it was severely affected, but now with links to the research institute who contacted me after they saw my products on the supermarket shelf, I am sure that I will receive better quality seeds for my next planting season,"* said Loy. *"I am in contact with the research team at Serere and so far, I have received material with information on the different varieties of groundnut which I find very useful,"* she added.

From the profit made from her groundnut business, Loy (Fig. 4.2) has been able to buy six goats which she rears for milk, she has rented a house for her children and herself, has been able to pay school fees for the children, and is able to also support some of her relatives.

**Fig. 4.2** Ms. Ademun Loy, a happy farmer and small-scale businesswoman, Soroti, Uganda (Photo: Manyasa E)

## 4.2    Against All Odds: Farmers Actualizing Their Dreams with New Groundnut Varieties

The Teso war deterred Mr. Martin Ocung's dream of completing high school and subjected him to poverty after the death of his two parents. He dropped out of Primary 7 due to lack of school fees, but as fate would have it, he met Dr. David Kalule, a groundnut breeder at NaSARRI and that was the beginning of Martin's journey with improved groundnut varieties. The 30-year-old Martin (Fig. 4.3 and 4.4) hails from Okulonyo village in Serere District, Eastern Uganda. He started planting the improved variety of groundnut in 2014.

Martin is different from the other farmers in his village as he works closely with the NaSARRI team to produce both breeder and foundation seed. He received lines from the research institute which were tried out in his farm. Martin also received additional training from the research team on how to properly manage his crop in the farm. According to Martin, the information disseminated to him has tremendously changed his life. *"I am wiser than I was 5 years ago, the new varieties I received from NaSARRI, perform much better than the old varieties I used to farm before,"* noted Martin. The seed class he produces has a higher market value than the other seed classes; this sees Martin making great profit.

The new varieties that Martin farms include Serenut 8 and 11. He works closely with NaSARRI on multiplying some new varieties that would be released in 2018. According to Martin, being involved in the Participatory Variety Selection is beneficial to him as he has first-hand information on the new varieties, and he can comfortably select the variety that has desirable characteristics.

*The climate has been very unpredictable in the last couple of seasons, so whenever I'm selecting a variety, I have to ensure that it is drought hardy, I go for varieties that are high yielding and pest tolerant. The old varieties are easily attacked by diseases like Rosette which highly affects the yield at the end of the season.*

Martins' life has changed since he got involved with the projects in 2013; not only has he increased groundnut production (from 1 acre in 2008 to 7 acres in 2016) he can now afford private schools for his five children and can take care of his wife's needs. In addition to that, he has bought more land; in 2005 he only had one acre of

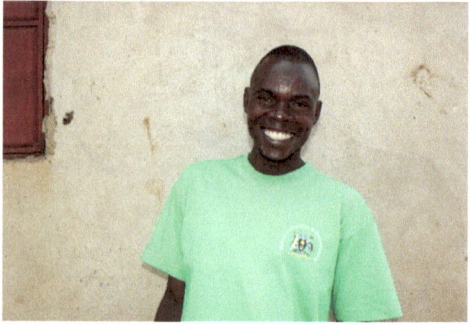

**Fig. 4.3** Mr. Martin Ocung, a farmer and former post war orphan in Serere, Uganda (Photo: Manyasa E)

**Fig. 4.4** Mr. Martin
Ocung's newly constructed
guest quarters (Photo:
Manyasa E)

**Fig. 4.5** Mr. Geoffery
Oceng with his bull bought
through proceeds from
improved groundnut
production, Dokolo,
Uganda (Photo: Manyasa E)

land, and from the money he accumulated from his groundnut business he bought 5 acres of land. *"Before my engagement with the Tropical Legumes projects, I got very low yield, in 2008 for instance I planted 192kilos of Eruduru Red groundnut and only managed to harvest 1600kilos, the yield was very discouraging and that's when I switched to the new varieties of groundnut,"* added Ocung.

Martin has even greater plans from the income he will receive from his groundnut farm, he has leased out more land and he is anticipating more yield which translates to more income for his household. Martin already owns a produce business, and from the money he will receive after the next season, he plans to build a bigger storage facility for his produce.

Mr. Geoffery Oceng's story is no different from Martins. Oceng (Fig. 4.5) survived on meagre earnings from his metal fabrication job. Poverty and bad grades couldn't allow him pursue higher education. He ventured into groundnut farming where he started off with the local varieties of groundnut which were readily available to him. From one acre of land he harvested 294 kg of unshelled groundnut. The results of his harvest weren't pleasant, but he was determined to be a great farmer.

> *I faced myriad challenges before I joined the Tropical Legumes projects I had no access to foundation seed, and I relied heavily on recycled ones. I was not able to differentiate between grain and seed when I went to buy seed from the market. I had no farming skills when I started out thus, I couldn't maintain quality of seed. In addition, I did not have any links to research centers.*

Through extension officers, Oceng has been fortunate to receive trainings on good agronomic practices which he has adhered to. He received improved groundnut varieties which he has been planting.

Geoffery (Fig. 4.6) earns 80% of his income from improved varieties of groundnut. From one acre he harvests up to 1050 kg of unshelled groundnut. His income has tripled, and he has invested the profit earned to improve his livelihood. He recently constructed a permanent house, purchased cattle, and started up a cloth business for his wife in the small town of Dokolo.

## 4.3   Scaling up Improved Groundnut Varieties: Tropical Legumes Revolutionized Groundnut Breeding in Uganda

A decade ago, groundnut was considered an orphan crop in Uganda; it didn't attract funding like most cereals and legumes did, and most of the people thought it would remain a subsistence crop. The tale has however changed, and groundnut are now the most important legume after bean in Uganda. The cultivated area of the legume is estimated at nearly 260,000 ha, representing 24.6% of the total arable land in Uganda. Since Tropical Legumes began in Uganda in 2011, the groundnut breeding program has become stronger than it used to be according to Dr. David Okello, a groundnut breeder at National Semi Arid Resources Research Institute (NaSARRI) in Serere.

Dr. Okello is the only groundnut breeder in Uganda and thus the task of ensuring developing new varieties of groundnut with desirable traits is solely on him. Groundnut are cultivated largely by poor farmers; this has enabled him work one on one with the locals in scaling up production of the legume. Groundnut consumption is also very popular in every part of Uganda.

From the funds received in the 5 years of working with the Tropical Legumes projects, a couple of things stand out for the breeding program. For starters, there has been an increase in production of breeder seed (below 1 ton to currently 5 tons in 2017) which in turn has increased the amount of foundation seed (over 80 tons in 2017) being produced by farmers in the country (Fig. 4.7). Dr. Paul Anguria, an Agronomist

**Fig. 4.6** Geoffery Oceng's newly constructed house in Dokolo District (Photo: Manyasa E)

**Fig. 4.7**   Groundnut nursery at NaSSARI in Serere, Uganda (Photo: Manyasa E)

at NaSARRI, notes that initially farmers were missing out on the new varieties and technologies, but since they got acquainted with them production has gone up.

*With support from the project, the farmers are now more informed when it comes to new varieties, diseases, maturity, types of soils and crop growth. Convincing them to take up the new varieties is however challenging as they are attached to the indigenous varieties. Through the project, the farmers have access to affordable and quality seed.*

Dr. Okello and his team at NaSARRI have been able to build a regional connection with different research institutions across and beyond Africa. The Institution, for instance, receives breeding material and technical support from CGIAR centers including ICRISAT Malawi, India, and the regional office in Nairobi. Before the project, the institution could handle 120 lines and approximately 40 crosses in a year. Overtime, with funding and support from Tropical Legumes projects and other partners, the team can comfortably handle over 600 lines and more than 60 crosses in a year. The lines and varieties have been shared with National Programs across Africa (Fig. 4.8).

*Participatory plant breeding*: The project also accords the research team to conduct participatory plant breeding; this enables the breeder to know exactly what the farmers and processors want in a variety and from that they come up with the appropriate variety. Conducting demonstrations and PVSs had been a hurdle previously due to financial constraints, but with funding more than 40 demonstrations is conducted in a year.

**Fig. 4.8** Groundnut
samples (Photo: Manyasa E)

*Digital data collection tool*: Data collection is no longer a challenge as the researchers have shifted to digital form of data collection. The Breeding Management Systems has ensured data is not lost or distorted. Tropical Legumes contribute to 40% of the research work on groundnut at Serere. Since the project's inception, the institution with funding from the project has been able to release nine new varieties of groundnut that are early maturing and more resilient to local diseases and drought, and hope to release more varieties with desirable variety traits in the future. Dr. Anguria Paul is hopeful that once aflatoxin and storage issues are addressed, ground-nut could easily overtake beans as the largest produced legume in Uganda.

## 5.1    Empowering Rural Women in Central Uganda to Achieve Higher Income and Improved Food Security with New Bean Varieties

Most smallholder farmers in Uganda often opt to farm the more popular crops which include coffee, plantain, cassava, sweet potatoes, and maize. Smallholder bean farming in Uganda is however overtaking the crops as farmers are seeing the potential that bean farming possesses. Women in the traditional Ugandan setting are considered as care givers and are tasked with ensuring food security. In Northern Uganda for instance, women are not entitled to inheriting land. When it comes to land sharing, women are only allowed to endorse the sharing. They are however given a portion where they can plant crops they can utilize in the household. The Tropical Legumes projects however strived to eradicate the stereotype that exists with the position of women in production for commercial purposes (Fig. 5.1).

The 50-year-old Marycian Nakaniako (Fig. 5.2–5.4) just like most farmers in Uganda previously utilized most of her land to farm Plantain, sweet bananas, coffee, and sweet potato; she however spared a little piece of land to plant yellow bean for household consumption. Nakaniako who hails from Kateera Village has noticed a significant improvement in yield with the improved varieties of beans she has adapted. Marycian, a seed producer, owns 5 acres of land but utilizes three acres for the production of improved bean. Before switching to the improved varieties, she only utilized 1 acre for bean farming.

The local varieties on her ¼ acre do not yield as much as her improved varieties; from 1 kg of old variety seed for instance, she harvests 7 kg as compared to the 20 kg she gets from 1 kg of improved varieties.

Selling the improved varieties seed is easy for Marycian as she is a member of Gomba Seed Producers' Cooperative which is a registered quality declared seed producing cooperative, that she joined in 2008. Before joining the cooperative, she sold her beans to merchants who did rounds in the village after harvesting season. *"I sold my beans as grain to local merchants who bought it at a very low cost; a kilo*

© The Author(s) 2020
E. Akpo et al., *Sowing Legume Seeds, Reaping Cash*,
https://doi.org/10.1007/978-981-15-0845-5_5

**Fig. 5.1** Bean farmers in Masaka District transporting their bean produce to the homestead after harvesting, Uganda (Photo: Manyasa E)

**Fig. 5.2** Ms. Marycian Nakaniako, a seed producer, owns 5 acres of land but utilizes three acres for the production of improved bean, Lwengo, Uganda (Photo: Manyasa E)

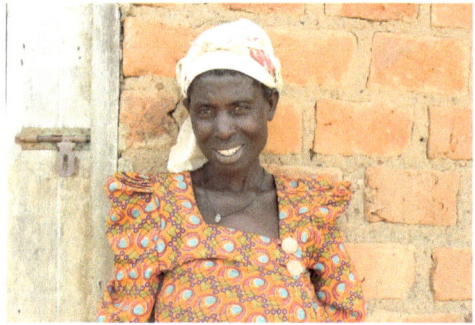

*of common bean is sold to merchants at 1500UGX,"* says Marycian. Seed production has however seen Marycian sells the same quantity of quality seed for 2000UGX.

With the additional income from her production of quality seed, Marycian has noted significant changes in her social and economic status. For her, the greatest achievements besides constructing a new house and purchasing more land is being able to educate her disabled daughter.

> *My eldest daughter is deaf, she wasn't fortunate enough to enroll in a special school when she was a kid; she dropped out at primary seven due to lack of school fees, but with the money I get from producing seed, I was able to enroll her for a tailoring course in Kampala. I even bought her a sewing machine which cost me 500,000UGX which she uses to earn a living for her family. Beans are the highest contributor to my household income, I also farm some maize, but it requires a generous size of land before I get the same amount of money I make from beans every season.*

**Fig. 5.3**  Ms. Marycian Nakaniako and her daughter Rebecca Namugenyi with the sewing she purchased for her after selling her beans, Uganda (Photo: Manyasa E)

**Fig. 5.4**  Ms. Amida Nabatanzi, a member of Kyazanga Farmers' Cooperative, Lwengo, Uganda (Photo: Manyasa E)

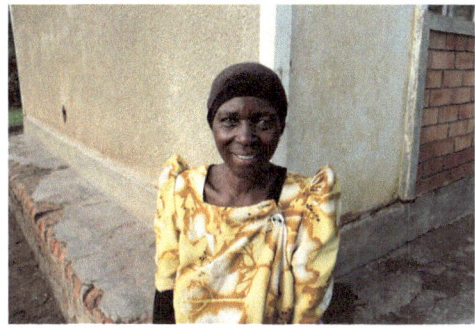

Ms. Amida Nabatanzi was married off at an early age, and after her separation from her husband, farming was the only source of income for her and her eight children. To make ends meet, she sold woven mats to her neighbors in Kalyamevuu Village in Lwengo District, Central Uganda, but the income was still insufficient to meet her family's needs. In 2015, impressed by how life had changed for her neighbors, she joined Kyazanga Farmers' Cooperative. Started in 1997 as a community group under Caritas Internationalis (MADO), the cooperative is a registered quality declared seed producer. *"I was required to pay a membership fee of 20,000UGX before enjoying the groups benefits. After I paid the fee, came the planting season and I was given 7 kg of improved seed (NABE 15) to test on my farm."* In addition to the improved variety seed, Amida received training on proper agricultural practices which include land preparation, line planting, spacing, crop management, and soil improvement.

Amida admits that the skills she obtained from the training were invaluable. In the past, the bean she planted produced about 80 kg, earning her less than 50,000UGX. In the season that she tested the improved bean variety, she harvested over 250 kg of bean. *"I barely got 80 kg before from bean farming but here I was getting more than 200 kg,"* noted Amida. *"I took my seed back to the cooperative which marketed it for me and paid me promptly."* In the second season she planted 20 kg of NABE 14 and harvested 900 kg.

The results from her first two seasons with the improved varieties of bean prompted Amida to increase her area of production; previously she utilized only ½ an acre for bean farming, but she increased it to 2 acres. Amida is among thousands of farmers who have benefited from the Tropical Legumes projects; the additional income from her bean production has enabled her to educate her children up to university level. She has also purchased two goats that she uses for milk production. She is looking forward to purchasing a television set and open a savings account to enable her to repair her house and build a store for her bean produce.

## 5.2  The Exciting Experience of Common Bean Seed Producer Cooperatives

### 5.2.1  The Successful Experience of Gomba Seed Producers' Cooperative, Gomba, Uganda

Gomba Seed Producers' Cooperative, formerly known as Gomba Seed Farmer Group, was established in 2006 by 25 farmers from villages in Gomba District, Central Uganda. The district receives minimal rainfall and thus the main economic activity is livestock. The group has over time grown and has registered 300 quality declared seed producers. The bean varieties grown by the group are NABE 2, 4, 15, 16, 2. The group procures seed from National Crop Resources Research Institute (NaCRRI) which they produce individually in their farms; they can sell the seed through seed companies and NGOs.

In a good season the group collectively produces over 200 tons of beans which is marketed together. The group is also the main supplier of beans to schools in Gomba and Mpigi Districts. They supply 50 tons of beans to schools in the two districts per season. Bean production has positively impacted the lives of many farmers in the groups. Most of them lived in grass-thatched houses but have now constructed permanent homes (Fig. 5.5). Some have bought bicycles, motorbikes, installed solar panels and bought electronics.

The group seeks to improve the livelihoods of community members and thus sometimes waive membership fees for members who can't afford and want to be part of the cooperative. These members are however loaned seed and the amount is recovered during sale. The group distributes quality declared seed to 90% of farmers in two subcounties where they are based. The sky is not the limit for Gomba seed cooperative as they have plans to build a bigger store due to their increased productivity every season. Plans are also underway to start seed processing and packaging for the group.

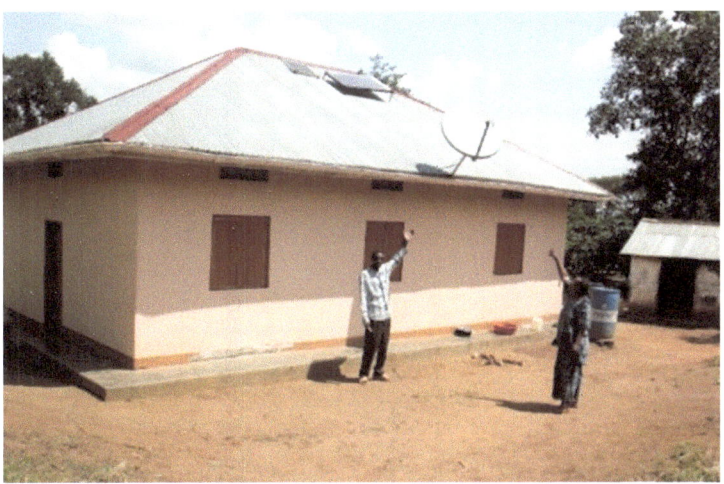

**Fig. 5.5** Mr. and Ms. Mutegana David, members of Gomba Seed Producers, showcase some of the assets they purchased from profit he got from selling beans, Uganda (Photo: Manyasa E)

Working as a stationery dealer in Kampala was not enough for 50-year-old Lameck Kyebonere to meet his wife's and children's needs. For years, he was often away from his family and the little profit he made catered for rent of his shop and his food and accommodation in the city. Mr. Lameck abandoned his business and returned to the village to focus on farming. According to Lameck, farming is the best decision he ever made, and he is not looking back. Through Gomba Farmers seed producers, he acquired NABE 4, an improved variety of bean. Lameck prefers the new varieties as opposed to the old ones because they are high yielding, tasty, climate friendly, and have high demand in the market. Lameck's productivity has increased from 800 kg in one-and-a-half acres to 2400 kg from the same land size. From the profit he made from beans he has upgraded his house and built a drying yard for his bean (Fig. 5.6).

The 51-year-old Lule Daniel from Kateera village is a trained quality seed producer from Gomba District, Central Uganda. Initially, he produced bean but sold it off to local merchants as grain. In 2006, Daniel came together with 24 farmers and formed Gomba Farmers Group. The group received training from NARO and Victoria seeds, and the trainings have been beneficial in improving the groups' bean production. *"I often used home saved seed during planting season, and I wasn't aware that the seed over time had become contaminated and susceptible to diseases and thus low yielding,"* explained Daniel.

The group was equipped with knowledge on land preparation, postharvest management, and pest and disease control. As a result of adhering to good agronomic practices, Daniel's productivity has tremendously improved; he previously managed to harvest 800 kg from two acres of land, but now from the same size of land, he harvests approximately 1.4 tons of improved bean. Daniel produces NABE 4.

Apart from beans, Daniel also farms maize, bananas, and coffee but beans contribute highest to his household income. He has built a permanent house and

**Fig. 5.6** From top left: Mr. Lameck and his family in front of his old house; top right: Lameck and his family in front of their newly constructed house. Bottom: Lameck on his drying yard, Uganda (Photo: Manyasa E)

purchased 8 acres of land from the extra income from beans (Fig. 5.7). His children have also been enrolled in high-quality schools in the District. He hopes to purchase a car in the coming year to enable him to move around the village efficiently as he was elected chairman of the group because of his industrious nature.

### 5.2.2    Improving Livelihoods of Smallholder Bean Farmers in Uganda: The Success of Kyazanga Bean Seed Farmer Cooperative

Formed in 1997 as a community group under Caritas Internationalis, Kyazanga Farmers' Cooperative Society Limited has grown to be among the most sought-after bean producers in Lwengo District. The group started with 26 members and a capital of 1,000,000UGX. Currently, the group has 1088 registered members and is valued at 200,000,000UGX.

The group aims at improving the livelihoods of peasant farmers in the district. Working with NARO accorded the group with the opportunity of testing lines which were later released to the public. The varieties released include NABE 1, 2, 3. The group has previously worked with other NGOs like AGRA to multiply improved

**Fig. 5.7**  Mr. Lule Daniel in front of his newly constructed house, Uganda (Photo: Manyasa E)

varieties. To enable proper management of the group, farmers have been organised in farmer groups; the cooperative collectively manages 60 farmer groups. Of the 1088 farmers, 400 are focused on seed production and 600 produce beans as grain.

To ensure high-quality seed is produced by the members, the group ensures that all the members receive adequate training on seed multiplication. They consequently conduct roadside demonstrations to create awareness and educate the community on the new varieties of beans available. This has contributed to high adoption of the improved varieties in the district. From the profit accumulated, the group has managed to employ qualified staff to help in managing the group. The group has also acquired assets and is putting on final touches to their new store which will hold more produce and new offices.

According to the groups' coordinator Joseph Asiime (Fig. 5.8), production of quality declared seed by the group is ensuring that farmers no longer access fake seed. They have also created a good market for beans for farmers due to their affiliation with seed companies and NGOs. The lives of farmers have improved drastically as the profit acquired from the production has enabled members educate their children, build homes, and some have even purchased motorbikes.

The cooperative hopes to register fully as a seed company to ensure they enjoy the full benefits of their hard work.

### 5.2.3  CEDO Ensuring Access of Improved Varieties of Bean to Smallholder Farmers and Vulnerable Groups in Uganda

Community Enterprises Development Organisation (CEDO) has played a significant role in converting beans to a commercial cash crop for the locality. The agro-enterprise development organization seeks to develop the production and marketing

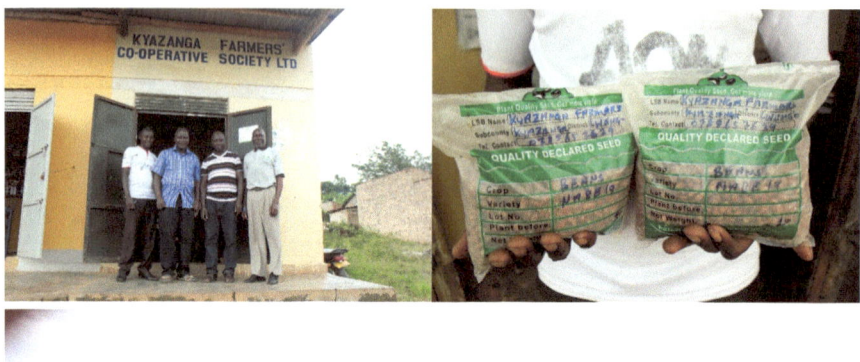

**Fig. 5.8** From top: From left Mr. Begumisa Stephen, Mr. Gabriel Luyima, Mr. Aaron, and Mr. Asiimwe Joseph in front of Kyazanga cooperative offices. Quality declared seed produced by Kyazanga Cooperative. Newly constructed offices and store for the Kyazanga Farmers' Cooperative, Uganda (Photo: Manyasa E)

capacity of local communities. Uganda has more than 32 seed companies which deal with a variety of crops. Five of the companies deal with beans only, CEDO being one of them. The company has tremendously improved from producing 900 kg of beans to more than 450 tons in a season.

CEDO was founded in 2000 with an aim to improve nutrition and household food security for smallholder farmers and marginalized groups in the society. Women, children, smallholder farmers, and commercial farmers were the target groups for the IRISH foundation formed group; HIV was rampant in the area at that time. After the IRISH foundation left, they supported the group and it was registered as a community-based organization. In 1998, after the cassava mosaic disease and the banana wilt, the farmers had no backstopping crop which a cash crop would be, having mainly focused on cassava and bananas. This facilitated the increase in bean production as most of the farmers resorted to bean farming. In 2000, the group became a community seed producing group. With links from food processors in Kampala, selling the beans was not a burden to the group.

Initially the group produced quality declared seed (QDS), but when they discovered the seed companies were making more money from just processing the seed they sold to them, they were motivated to register as a seed company. The group followed up with the Ministry of Agriculture and were linked with NARO who came and provided guidance and support on methods of producing seed. The International Centre for Tropical Agriculture (CIAT) provided learning materials and seed and training ideas on marketing. Farm trials and demonstration fields were conducted by the research team from NARO, where farmers would learn bean farming practices. The farmers were also part of the participatory variety selection (PVS). The research team at NARO has played a vital role in the increase in bean production for CEDO. NARO sensitized them on parameters to follow when producing seed, offered trainings on different processes and brought in stakeholders from different disciplines. The enterprise has also been directly engaged with staff from the Ministry of Agriculture who assist them with inspection and standardization of their produced seed.

The enterprise started with two varieties in 2000 and as of 2017 they were producing more than ten varieties of beans. *"Before 2007 we majored in one district, but we have now increased to 5 districts. We currently supply beans to 11 districts in Greater Masaka, and work more with women in bean production than men."* The group chairperson narrated.

## 5.3   Strengthening the Bean Breeding Program and Value Chain in Uganda

The Tropical Legumes projects majorly aimed at enhancing capacity in the breeding sector. A lot of emphasis has been put in developing new materials from already existing germplasm. The bean breeding program work closely CIAT's program. CIAT owns the largest bean gene bank in Cali, Colombia.

> Before 2012, there was hardly a bean seed system to talk about in Uganda, but this has changed over time. Tropical legumes contribute 30% to the research program at NaCCRI. Initially the breeding program was haphazard," said Dr. Stanley Nkalubo.

Fig. 5.9). *"We have moved from incorporating things randomly to straightening out breeding pipelines policy."*

Digitization of data collection which was facilitated by acquisition of tablets with funds from Tropical Legumes projects has ensured data is not lost or distorted by the field technicians; they have also been adequately trained on how to use the gadgets. NaCRRI could only handle two traits at a time when it came to breeding new material but currently we handle close to ten traits at the same time. Annually the institution produces approximately 30,000 kg of breeder seed.

**Fig. 5.9** Dr. Stanley Nkalubo in bean nursery at NaCCRI in Namulonge Uganda (Photo: Manyasa E)

Fake seed was a widespread problem in Uganda before the project started in Uganda. Most farmers did not have the capability to distinguish between fake seed and quality seed but over time, with trainings and being involved in the breeding process, the farmers are more knowledgeable. The seed system has also seen growth in seed companies as the program is working closely with seed companies (Pearl seeds and agro dealers in Uganda) to multiply the beans and ensure that farmers can have easy access to quality certified seed. Community seed multiplication has enabled farmers to easily access quality seed. The number of registered seed companies involved in bean production and distribution has gone up from 7 to 17. The 17 companies work closely with the research institute at Namulonge.

Through the research institute, more farmers have been reached by extension officers. TL is keen to empower women in their projects, and over the years, women involvement in bean production has been notable. Gabriel Luyima, a researcher at NaCRRI, notes that when women are involved in production there is less wasteland; this translates to more income for the family. Like in many households in Africa, the woman ensures food security for the family, and thus when involved, she will ensure that there is enough produce to feed the household and some for sale. Through sensitization, the women in Uganda have a different perception on beans; before it was merely for consumption, but it since has changed and they now look at it as a commercial commodity. The women now have a voice of their own, are more confident, and have gone a notch higher to even speak at major meetings (e.g. conferences) on the importance of bean production. The yield gaps between men and woman has also significantly reduced over time.

Benchmarking activities between farmers which were organized by the project has made them more exposed and knowledgeable. They have in addition learnt the nutritional benefits of the improved bean varieties. The yield has gone up for both the male and female since the inception of the project; pregnant women now utilize beans as supplements for zinc and iron.

A total of 13 varieties have been released and promoted through the Tropical Legumes projects. NABE 15 for instance was released in 2010 through Tropical Legumes II, the variety has been up-scaled to South Sudan and Northern Uganda. Two more varieties which are high iron and zinc are in the pipeline and will be released in due course.

## 6.1 Enriching Lives of Women: Groundnut Production and Processing Is a Mine of Wealth for Women in Northern Nigeria

Ms. Hadja Talatu Idrissa (Fig. 6.1), a community women leader, is the chairperson of the Bunkure women group that is active in groundnut production and oil processing. The group which counts 25 members joined the TL projects' family 4 years ago. They started growing a small seed pack of 5 kg in their community farmland. *"It was the harvest of this seed pack that we revolved and planted in a bigger farm plot in the following year,"* reported. Hadja.

On 1 ha plot, the Bunkure women harvested a total 25 bags of the improved variety SAMNUT 24 against 13 bags they got from 1 ha plot with the old variety.

In addition, the group made more money out of the haulms of the improved variety SAMNUT 24. *"We sold the haulms of the improved variety up to 30 000 Naira against 12 000 Naira of the local variety,"* says Idrissa (1USD equals 360 Nigerian Naira during that period). *"The improved variety, SAMNUT 24, has higher haulms yield and is much appreciated for animal feeding because of its taste and digestibility which is better,"* she adds.

In 2017, the Bunkure women group produced about 3.5 tons groundnut. The grain is used for household consumption, while the groundnut haulms are sold and the money was used to start dry season groundnut production in 2018. *"We don't sell our grain produced, rather, we keep it and process part of it into oil and many by-products which we further sell. Apart from money made from the processing activities, individual members contribute 200 Naira on a weekly basis for the savings box. A weekly savings of about 5000 Naira is kept into the group's bank account."*

The interest from this saving permitted the group to conduct many activities to help the community as a whole, including restoring the community health center and its primary school. *"We use part of our savings to clean up the community health centre and pay for basic products to sustain the centre. The hospital is now cleaner than before and offers heal thier working environment to staff and to*

E. Akpo et al., *Sowing Legume Seeds, Reaping Cash*,
https://doi.org/10.1007/978-981-15-0845-5_6

**Fig. 6.1**  Ms. Hadja Talatu Idrissa, leader of the Bunkure women group, producing groundnut and processing oil, Nigeria (Photo: Diama A)

*the patients. Before, people were afraid of visiting the hospital because it was in a bad condition which would expose them to other infections caused by insalubrity of the hospital rather than getting treatment. Also, the compound of the hospital was so weedy, that the nurses could not stay overnight as they were afraid of many animal attacks. Now that we have cleaned up the hospital, they are no longer afraid of staying for long hours in the hospital. As a matter of fact, the health centre now offers 24 hours full services and the nurses are ready to attend the patients at any time, day and night."* The women group contributed to repair the beds in the hospital, and this offered a more convenient place for admitted patients including pregnant women. The group also contributed to restoring the doors and windows of the community's primary school.

Ms. Hadja Talatu (Fig. 6.2) says that the group has contributed to improving the education of children within the community of Bunkure. *"Before, most of our children used to school up to primary level. Now, we have children at Universities in capital cities."* At a more personal level, Ms. Hadja Talatu says she was able to attend the pilgrimage in Mecca, Saudi Arabia, and is proudly happy to say how much progress the women group was able to make from groundnut grain production and processing as well as from groundnut haulms business. *"I have a lawyer, a doctor and even an agricultural extension worker."*

In a country where women access to land is still a major issue, Ms. Hadja Talatu and many other members of the Bunkure women group are now happy owners of farm lands and many working bulls.

Happiness have different meanings to all and Hadja Talatu (Fig. 6.3) together with the Bunkure women group seems to have reached their goal. Tropical Legumes projects have put a smile on their faces which they have gratefully translated into their community in many ways. The group was recognized in 2015 by the State

**Fig. 6.2** Ms. Hadja Talatu Idrissa showing the legal document of the creation of the Bunkure women group in 2008 and a recognition from the Governor of Kano State for their contribution to the development of the community, Nigeria (Photo: Diama A)

**Fig. 6.3** From left to right: Ms. Hadja Talatu Idrissa and Ms. Afsat (Agricultural Extension Agent of the Bunkure Women group, Northern Nigeria) (Photo: Diama A)

Governor for their substantial contribution to the development of the entire community.

## 6.2    Seed Companies Helping to Lift People Out of Poverty: A Breakthrough in Seed Systems with Maina Seeds and Greenspore

Maina Seeds Ltd. and Greenspore Ltd. are two seed companies based in Kano, Northern Nigeria. Both seed companies have an equal partnership and hence benefit equally from the Tropical Legumes projects, specifically groundnut and cowpea seed production. Tropical Legumes projects (I-II-III) and other various projects have created awareness and raised capacities along the value groundnut and cowpea value chains, including seed companies.

> *Many farmers were afraid of embracing cowpea production because of the Maruca infesta-tion which caused closer to total loss in the fields and made the farmers poorer,* says the late Awalu Balarabe, the Managing Director and Chief Executive Officer of the Maina Seed Limited company (

Fig. 6.4). *The project came with technologies for Maruca control along with prom-ised varieties that are higher yielding, relatively resistant to some biotic and abiotic stresses. It introduced improved storage bags for cowpea which helped a long way. Also, extension services were provided by our Agricultural Development Program*

**Fig. 6.4** Late Awalu Balarabe, the Managing Director and Chief Executive Officer of the Maina Seed Limited Company, Nigeria (Photo: Diama A)

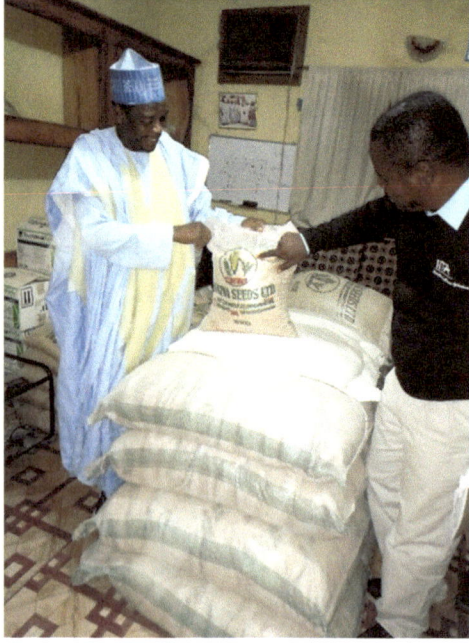

*(ADP). In the seed industry, we felt comfortable because we knew that the breeders did not abandon us. We felt comfortable in going into larger production and getting more seed out-growers,* he explained.

Improved varieties are now made accessible and sold out to farmers using small seed pack of 1, 2, and 5 kg. Another benefit is the improved linkages with more agro-dealers and seed out-growers into the seed value chain and the fact that several farmers attached to the project were linked to seed companies and have become immediate sources of certified seed for seed companies.

Before the projects, farmers were cultivating cowpea and groundnut which were limited to their personal uses. The TL projects came with improved varieties which saved farmers from cropping the same varieties years after years, with poor yield. Farmers were convinced with demonstration plots and accepted new technologies, and by doing so they were able to increase their yields. *"It gave reasons to the seed companies to go into mass production of this cowpea. Maina Seeds used to produce less than 2 tons of cowpea per year. Since 2015, before my engagement with the Tropical Legumes projects, I got very low yield, in 2008 for instance I planted 192 kg of RUDD Red groundnut and only managed to harvest 1600 kg, the yield was very discouraging and that's when I switched to the new varieties of groundnut. We produced up to 30 metrics tons. In years 2016 and 2017 Maina Seeds has seen production reach up to 40 metrics tons of cowpea,"* adds the late Awalu.

In the same way, new groundnut varieties (SAMNUT 24, SAMNUT 25, and SAMNUT 26) were promoted through demonstration plots and field days were conducted over the years in farmers' fields which they adopted. *"For the first time through the project, we have access to breeder seed of improved groundnut varieties,"* says the late Awalu.

Such support including trainings on seed production techniques of both cowpea and groundnut, quality control including aflatoxin pre- and postharvest management support, were given to seed companies. Other capacity building focused on business skills and seed entrepreneurship *"We have also scaled down such trainings to our out-growers,"* adds the late Awalu.

Greenspore Agri Limited is a seed company based in Kano State operating in eight states in Nigeria. The company started with groundnut and cowpea seeds at a very low level, because primarily there were no sufficient foundation seeds for these crops and there was need for promotion of these crops. *"We are so lucky to have become collaborators in the Tropical Legumes projects II and III,"* notes Mr. Shehu (Fig. 6.5). *"We didn't have enough access to foundation seed, with less breeder seed in stock. More foundation seed and breeder seed became available and gave us an advantage of extrapolating and increasing our production and yield. We have at least 3 new varieties and the farmers have been very happy because most of them are resistant to leaf diseases, including rosette. With regards to cowpea production, greenspore started with a little amount of foundation seed that was available. Now that we have more foundation seed available, we can produce more certified seed."*

As the availability of the foundation seed increased, the seed companies were able to promote them among a network of out-growers, thus improving their revenue and capital. A good case study is the Maina Seeds Company which according to

**Fig. 6.5** Mr. Balarabe Shehu is the Managing Director and Executive Director of Greenspore Agri Limited, Nigeria (Photo: Diama A)

the Managing Director, the late Awalu Balarabe, the capital increased from 1 million naira to 250 million naira in 2017. *"When we started, we had no office to call our own, today we have established a permanent office, a showroom, a shop, an agricultural laboratory, and a conference room. We were using laborers and horses to cultivate our farm; today, by the grace of God, Maina Seeds Company has its own tractors, a combined harvester and threshers. We were enhanced and definitely this project has helped,"* testifies the late Managing Director of Maina Seeds.

> *"Our biggest success with Tropical Legumes projects is the introduction of the company to scientists and to extensionists. They have exposed us to many other players with several best practices in the seed industry. The benefits are many and beyond mere finances. We have been enhanced as a company,"* the late Awalu concludes.

Long ago Nigeria was known for famous groundnut pyramids. The country is also among top producers and consumers of cowpea worldwide. However, the market demand is very high and seed companies are still making great efforts to meet the demand. *"Although breeder seed is now available in the system, it is not enough,"* says the Managing Director of Maina Seeds. *"We need more of breeder and foundation seeds."* the late Awalu added that despite the company is increasing its production to 30 metrics tons, he is still not able to meet the demand of the market. *"With the local demand being high, there is need for a concerted effort to increase the delivery of breeder and foundation seed to the seed companies, so that we will be able to produce more, sell more to farmers and reach out to other states outside our immediate operating states."*

*"If seed companies continue to get the right support in getting breeder seed and our scientists get more support to produce more breeder seed and foundation seed, the seed companies will scale it up and farmers will take in large quantities,"*—believes the late Awalu.

Efforts are being made to bring the Nigerian groundnut pyramids back and farmers are being encouraged by the new SAMNUT 24, SAMNUT 25, and SAMNUT 26 which are high yielding. Also, millers sprinkled, and the food industry is using the produce to make various snack.

## 6.3   New Groundnut Varieties Released by Breeding Program After Decades

*"The improved groundnut varieties SAMNUT 24, SAMNUT 25 and SAMNUT 26 released to farmers between 2012 and 2013 are our proud achievements from Tropical Legumes projects. Three new lines are being nominated to the National Variety Release Committee of Nigeria for registration and release in 2018,"* says Prof. Candidus Echewku, groundnut breeder, Institute of Agricultural Research, Nigeria.

## 7.1  Woman Transforming Her Family Life and That of Her Community

Ms. Hadja Salame Shaibu (Fig. 7.1) is a cowpea producer and processor in the Local Government Area of Dawakin Tofa. She grew up in a farming community where crop- livestock integration is a tradition. When she got married few decades ago, Salame continued farming and keeping livestock to support her new family, mainly in sorghum and cowpea production.

Few years ago, Salame also started processing cowpea into local dishes (Moi-moi, Accra, Danwake, and many other products). She also cooked and sold a special dish made from a combination of pasta and cowpea which was well appreciated by the consumers and made her successful.

It all started 2 years ago when Salame met with the project extension agents and was given a small seed pack of a new improved cowpea variety for testing. *"I used to produce 2 bags (200 Kg) of the local variety in a cropping season but with the improved variety, I harvested up to 5 bags (500 Kg),"* says Salame. *"After a farming season, I process my produce into various products and sometimes, I buy additional cowpea of 400 to 500 Kg for more processing. In a year, I can process up to 1,000 Kg of cowpea."*

With the new improved varieties that are higher yielding, Salame explains that she could make more profit from her business. *"The new varieties produce more flour and taste better. I think it has improved nutrition within the community where many people were lacking some essential proteins in their daily diet. Now that cowpea has improved nutrition, it has helped me improve my incomes,"* she happily adds.

As a matter of fact, Salame has now expanded her business and is now the owner of a new car. *"I have bought two bulls to help with farming activities; I am building a new house made of concrete blocks; and I have purchased a new car that my son is now using for commercial purpose and to transport goods. Also, I am taking care*

E. Akpo et al., *Sowing Legume Seeds, Reaping Cash*,
https://doi.org/10.1007/978-981-15-0845-5_7

**Fig. 7.1** On second left, Ms. Salame and family in front of their old house and right (Ms Salame—seen third from right) in front of their new house along with their new car. Also seen are Ms. Salame's husband (fourth right) and her son (second from left—with his hand on the car), Nigeria (Photo: Diama A)

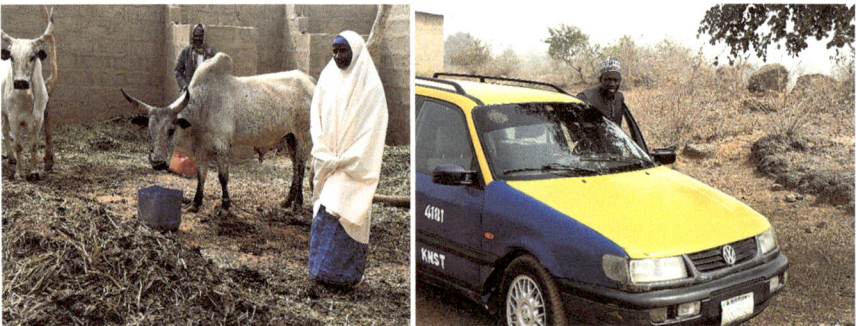

**Fig. 7.2** On left, Salame and husband with the bulls and on right, Salame's new car used by her son for commercial purposes, Nigeria (Photo: Diama A)

*of everything in the house as I don't wait for my husband to pay for everything,"* she adds.

More recently, Salame increased her farm size and has money to pay laborers unlike before when she used to do all the work manually by herself.

Salame started with small seed packs of improved variety, growing later to producing on 2 ha. The first benefits were used to buy small ruminants and bulls which she later disposed and bought a car and concrete blocks for building a house. With average incomes of 500,000 Naira per year, Salame who has been in cowpea production and processing for 10 years says improved varieties have boosted her activities in the 2 past years. She is now the only woman who owns a car in the village (Fig. 7.2). This has inspired more women and created awareness in the entire community for a growing interest into cowpea production. Cowpea production has exposed Salame not only within the community but also to the local government area as a whole.

## 7.2  From Grass to Great: Tropical Legumes Projects Have Changed the Pattern of Agricultural Extension in Northern Nigeria and Nigerian Farmers Increase Their Cowpea Production

Cowpea remains vital for many smallholders in Nigeria where it is grown primarily for human consumption. Also, the fodder market of the crop has witnessed a considerable success in the animal feeding growing market during the past years. The Dawanu market in Kano (Fig. 7.3), Northern Nigeria, is the largest cowpea market in the world. In Nigeria, the Tropical Legumes projects have increased the adoption and uptake of improved cowpea varieties by farmers in Northern Nigeria. As a result, seed production and supply of improved and farmers' preferred varieties were significantly enhanced. Between 2007 and 2013, more than 530,000 tons of certified seed (CS) and quality declared seed (QDS) were produced in project target zones in Northern Nigeria.

As result of capacity building of the national breeding system in the Institute for Agricultural Research (IAR), 4–10 tons of breeder seed were annually injected into the groundnut seed value chain to meet the national demand against 500–1000 kg produced prior to the project interventions.

The role of the agricultural extension agents was crucial in the promotion of these improved varieties according to Mr. Sayi Ado Oumar (Fig. 7.4), an extension agent working with nine (09) communities in the Local Government Area (LGA) of Tsanyawa, Kano State Agricultural and Rural Development Authority (KNARDA). He recalled how much Tropical Legumes projects, especially in its third phase of implementation, was key in changing the extension pattern in Northern Nigeria.

**Fig. 7.3**  A scene in Dawanu grain market, Northern Nigeria (Photo: Diama A)

**Fig. 7.4**  Mr. Sani Ado Oumar was able to buy a new car from the benefits of his product, Nigeria (Photo: Diama A)

*Before TL, our institution used to provide us with pre-season training. TL has strengthened these efforts with a more targeted trainings and capacity building of extension agents in agronomic practices, pest management, safe and effective use of pesticides, data collection and record keeping, post-harvest management including cowpea storage as well business and marketing of agricultural products. The impact was tremendous, says Mr. Ado.*

According to Ado, the greatest impact of a large uptake of technologies was achieved because of the trainings offered to all extension agents within the local government area. *"I have always provided a training to fellows who did not benefit directly from the project trainings. It has changed the pattern of extension activities in the Local Government Area of Tsanyawa. The training made a change in me which I was able to translate into 28 other extensions workers in the nine communities covered by the project in our Local Government Area."*

The project capacitated the extension agents with means for transport, and thus increased their mobility and facilitated a closer contact and monitoring of outreach activities. *"We started with few farmers in 2015 and with time, farmers took interest into cowpea production and wider acceptance and adoption rate has been registered due to our extension work. Trust has increased in new varieties and farmers have more confidence in using improved agronomics practices,"* explains Mr. Ado.

On a more personal level, Mr. Ado says that TL projects were a huge opportunity for development and progress. *"This project has enabled me to start my own cowpea production farm. Providing training to farmers has motivated me to embrace cowpea seed production. The project inspired me to create and register my own seed company, Ausye Agro-chemicals and seed company Nigeria Ltd,"* says Ado who is now the owner of 2 ha where he produces cowpea. *"It has changed me from grass to great; an extension worker has started from grass to great due to TL projects,"* he

**Fig. 7.5** Mr. Sani Musa, cowpea producer, Nigeria (Photo: Diama A)

adds. Mr. Ado was not only able to improve his revenues, but he was able to invest in a new car to improve his mobility in the field and reach out to more farmers.

With the support of the projects, several technologies that have consumer-preferred traits were developed and released. These improved technologies include newly released cowpea varieties that are high yielding and fast maturing with resistance to some of the major diseases, pests, nematodes, parasitic weeds and adapted to sole planting or intercropping. The cowpea varieties have increased the interest of farmers to convert to cowpea production.

The success of Mr. Ado is not an exception in Tropical Legumes projects' interventions zones in Nigeria. Mrs. Samale Shaibu from Tsanyawa Local Government Area has a fairy tale. *"With the improved varieties, I produce up to 45 bags per season which I then process into various products including Danwake, a local dish which is well appreciated by the consumers. From the sales, I bought small ruminants, two bulls, and a commercial bus. I am now building a new house with concrete blocks in my village."*

Mr. Sanu Musa (Fig. 7.5) from Bagadawa Local Government Area is not new in cowpea production, but he started a new experience with the crop in 2017, when he cultivated his first improved variety of cowpea. *"I harvested 14 bags where I could barely get 3 bags with the local variety. I sold 13 bags and used the earnings to build a house. I can pay school fees for my children and have improved their clothing as well other enjoyment. Many fellows have witnessed my success and are willing to start cowpea production in 2018."* Musa who appreciates the improved varieties of cowpea says he hopes that the project will continue supporting farmers.

*In 2017, I built a house; in the coming year I hope, I wish, and I am willing to construct 3 additional houses for the comfort of my family,* Mr. Sanu Musa concludes.

## 7.3   Enhanced Production of Early Generation Seed of Cowpea

*The quantity of breeder seed of improved cowpea was raised up to 4-10 tons annually to meet the national demand. Before, only a few kilograms (500 to 1000) of different varieties were available in the system,* says Prof. Abubakar, cowpea breeder, Institute of Agricultural Research, Nigeria.

# Better-off Women Boosting Groundnut Business in Ghana

## 8.1    More Groundnut Means Shelling Is a Business

Groundnut was one of the biggest breeding programs in Ghana in the mid-nineties, but the production declined because of many factors including the rosette disease and the fact that there was no dedicated breeder of groundnut for over 10 years. According to Dr. Roger Kanton, Deputy Director of CSIR-SARI (Council for Scientific and Industrial Research - Savanna Agricultural Research Institute), it was then, in 2015, with the support of the Tropical Legumes Projects that the groundnut breeding program was reinitiated. *"Only a few local germplasms were available,"* adds Dr. Richard Oteng-Frimpong, a young groundnut breeder, who came along with the support of the Tropical Legumes projects to start again the breeding program in 2015.

*Groundnut production and processing in* Nyankpala, Northern Ghana, *is now seen as a business.* Umar Jibril, a fabricator of groundnut shellers, narrates, *"In 2006, we could barely fabricate one or two groundnut shellers in the year. Now we fabricate up to 4 groundnut shellers per month; the demand is very high to a point that clients must place an order well in advance. Our clients used to be the villagers but nowadays our clientele is made of small and medium enterprises."*

Since 2015 and through the Tropical Legumes projects, SARI collected more than 300 new materials mainly from ICRISAT. Up to 109 breeding lines were developed and with the assistance of the projects new irrigation facilities were developed and two advanced generations were created each year. Main traits including resistance to foliar diseases, drought tolerance, oleate, early maturity, and oil content were investigated, tested, and evaluated with farmers. As a result, a total of five varieties were promoted and 3 tons and 8.4 tons of foundation seed were made available in 2015 and 2016, respectively. About 42 tons of certified seed were produced in 2016. According to the Deputy Director of CSIR-SARI, three new varieties are in the final stage to be released. *"The material release will bring SARI and Ghana back to groundnut breeding,"* explains Dr. Kanton.

The existing old varieties had a yield potential of 1.8 tons/ha while the new materials have a yield advantage over the existing varieties between 10 and 30%. Before

E. Akpo et al., *Sowing Legume Seeds, Reaping Cash*,
https://doi.org/10.1007/978-981-15-0845-5_8

the Tropical Legumes projects, the amount of breeder seed produced per year was less than 0.2 ton. With the support of the projects, 2 tons were produced in 2015 and 6 tons in 2016.

*"The increase in breeder seed production means that seed companies have better access to breeder seed to produce foundation seed; it also means that out-growers are producing certified seed and that farmers can enhance their community seed production,"* says Mr. El Hadj Abdul Razak, Director of Heritage Seed Company. Since 2016, the company was linked to 300 nucleus farmers that he supplies with foundation seeds for the production of certified seed. The company produces on demand and supply to many partners including NGOs. *"The market is very interested in new varieties and for 3 years now, we have been having a problem in meeting the demand,"* he says.

> *From zero, 3 tons and 8.4 tons of foundation seed were produced in 2015 and 2016 respectively. A total of 42 tons of certified and Quality Declared Seed (QDS) were produced in 2016,* says Razak.

This success is the results of several trainings organized on good agronomic practices among groundnut producers and key players along the value chain within multi-stakeholder platforms (MSPs) established under TL Projects. Such trainings focused on groundnut seed production and agronomy, weed management in seed fields, pest and diseases management, post-harvest handling, establishing community-based seed systems and seed certification. Also, trainings were provided on maintaining a viable seed business targeting groundnut seed producers, out-growers, and inputs dealers. From 2015 to 2017, more than 8000 farmers were reached with improved varieties.

With more breeder seed, foundation seed, certified seed, and Quality Declared Seed (community seed systems), not only farmers, out-growers, and seed companies could increase their production and yield but a range of new businesses also appeared in the value chain, including shellers that would speed up the processing. Mr. Umar Jibril (Fig. 8.1) is a sheller maker in Nyankpala, Northern Ghana. He has noticed a boom in the demand for groundnut sheller during the past 2 years. *"In 2006, we could fabricate barely one or two shellers in the year, but we can now fabricate up to 4 shellers per month as the demand is very high to a point that one must place an order well in advance,"* says Jibril. *"Initially, our clients used to be the villagers, but with time our clientele has diversified."* The fabricator who used to work alone with his uncle in a small garage is now the owner of an enterprise of ten employees.

One sheller can be sold up to 3600 Ghana Cedi (1 USD equals 4.5 Ghana Cedi during that period), and one machine can shell up to 4000 kg/day. Each bag of 40 kg is shelled at a cost of 3 Ghana Cedi. *"People had seen groundnut production and processing as a business,"* says Jibril. *"Now we have many entrepreneurs who buy our shellers to make them work in village"*. Figure 8.2 shows active women operating at a groundnut shelling station.

With the increase in groundnut production, Mr. Seidu Bushira and his wife Seidu Andani (Fig. 8.3) have initiated a groundnut shelling business near Tamale city. The

**Fig. 8.1**  Mr. Umar Jibril in front of two fabricated shellers, Ghana (Photo: Diama A)

**Fig. 8.2**  On left: women operating the groundnut sheller. On right, a woman prepares shelled groundnut for the market, Ghana (Photo: Diama A)

couple which employs about 20 daily workers has two groundnut shellers. *"Before the boom in the groundnut production, we use to shell only 400 kg per day, now we can process up to 600 kg,"* said Bushira. Groundnut producer like Mr. Abdul Majeed can now shell their groundnut as fast as possible to proceed to the market. Mr. Abdul Majeed (Fig. 8.4), a diploma holder in Business studies, has taken up the existing market opportunity to farm 4 acres on which he harvested 2.5 tons. With money earned, the young student plans to pursue a higher national diploma in Business studies with management option.

**Fig. 8.3**  Mr. Seidu Bushira (first right) and his wife Seidu Andani (second left) (Photo: Diama A)

Women in the communities are heavily involved in groundnut business (Fig. 8.5)

## 8.2    Village Savings and Loans Association (VSLA) Boosting Groundnut Production Among Small Holder Farmers in Northern Ghana

Empowering small holder farmers to be financially independent is very crucial to ensuring a constant crop production system. Access to credit in the form of cash and inputs for production has always been a major constraint bedeviling the crop production sector. Interest rates charged on credit accessed from financial institutions tend to be often high, thus making farmers reluctant in accessing them. Another major challenge farmer's face in accessing credit is the repayment terms of credits. Financial institutions have often had hectic time dealing with farmers when it comes to recovery of the loan farmer received for

**Fig. 8.4** First on left, Mr. Abdul Majeed (a student and groundnut producer who has brought his produce (2.5 tons) for shelling (Photo: Diama A)

**Fig. 8.5** Groundnut businesswomen in communities

crop production. This is because most of the farmers are not able to repay back their loans.

Several organizations have often rolled out programs aimed at reducing the plight farmers go through in accessing credit facilities by providing inputs while farmers repay back with grain, but this has yielded very little results. This is because farmers refuse to repay credit incentives packages given to them, making it difficult for them to roll out their interventions to other communities or individuals who need assistance.

In Northern Ghana, TL project has partnered with SEND-Ghana (Social Enterprise Development), an NGO to put in place a Village Savings and Loan Association (VSLA) in five districts of Ghana (three in northern regions, one in Upper West Region and another one in Upper East region). Since then, VSLA has been used as a platform to help groups raise funds to support activities that require financial assistance (Figs. 8.6–8.8). *"This VSLA is a self-help initiative, where group members come together to raise funds through weekly or monthly contributions within a given period of time,"* explains Mr. Desmond Adogoba, Gender and Social Scientist, SARI.

### 8.2.1　Formation of Village Savings and Loans Association (VSLA)

The implementation of the concept was derived from strategies developed from a gender workshop organized by the TL projects, aimed at bridging production gap between smallholder male and female farmers. *"The objective of this initiative is to give members the chance to save money that will be used for groundnut seed*

**Fig. 8.6** TL improves women access to credit and strengthens their participation in groundnut seed systems, Ghana (Photo: Diama A)

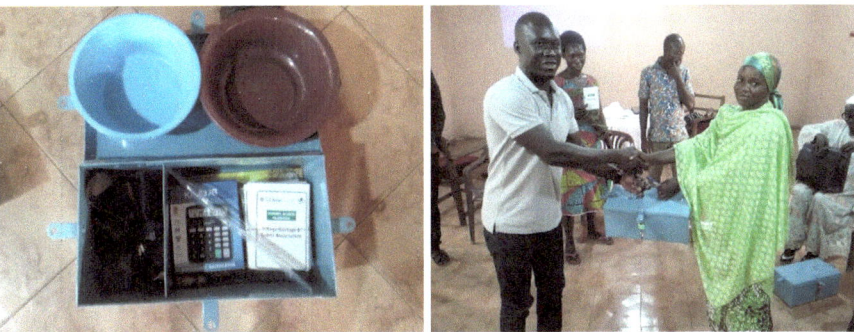

**Fig. 8.7** On left: The complete VSLA kit is composed of a calculator, a membership card, a metal box for keeping the savings, and two plastic boxes used to collect money during weekend meetings. On right: Presentation of VSLA kits to community volunteers (from left to right Mr. Desmond Adogoba (Gender and Social Scientist, SARI) presenting a VSLA box to a community volunteer.)

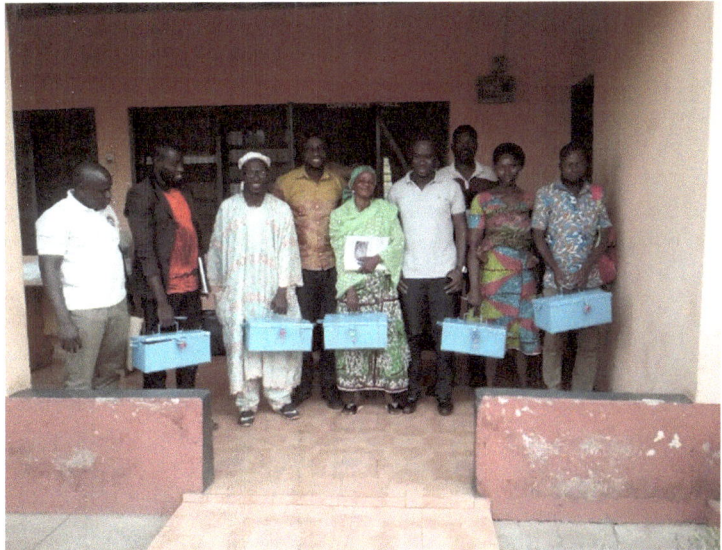

**Fig. 8.8** Group picture with VSLA community volunteers in January 2018 (third from left is Mr. Sardi Linus Handua, Secretary of the Gbimsi "Tilanngum" VSLA, Ghana (Photo: Diama A)

*production in their respective communities as well as support other households in on-farm activities which would have required borrowing money from external sources to execute them. This initiative helps equip members to be financially independent and strengthen the groundnut seed production system at the community level,"* said Mr. Desmond Adogoba. *"We are not only empowering women, we are strengthening and engaging them into the seed systems,"* he adds.

The VSLA initiative is piloted under the projects in five communities across five districts in northern Ghana (the Wantugu community in the Tolon District, the Salankpang community in the Mion District, and the Gbimsi community of the West Mamprusi District) and each VSLA has 150 members across the five pilot communities.

Mrs. Patience Ayamba is the program coordinator for the SEND-Ghana Livelihood and Food Security program. She is based in the Salaga Office of SEND-Ghana, where she coordinates activities of the SEND-Ghana livelihoods program across northern Ghana. She said that partnering with SARI through TL projects has helped them in expanding their gender training activities and VSLA into Northern Ghana. She expects that this partnership will lead into taking these communities to the Gender Family Model Concept where a husband, wife, and children are included. As she says, *"VSLA is just a part of the Gender Family Model where both men and women get to understand their roles in the families. We have seen women taking more participation in decision making at the family level and even at the community level; with more women taking leadership role. With this model, we have seen men who are willing to support their wives in household burden, paying more attention to children's health needs and women have seen their views more respected by their husbands. We have seen these things happenings with SEND-Ghana fostering project. With lot of hard work, we expect to get the same good impact from five communities in which we are collaborating with the SARI and TL projects."*

The formation of VSLA will be used to address three major problems identified during a study conducted to determine gender productivity gaps among small holder farmers across northern Ghana under the TL projects. These are access to credit, access to improved variety seeds, and access to fertile lands.

## 8.2.2    Training of VSLA Community Volunteers

One community volunteer was selected from each of the five VSLAs for a special training on the management plan of the VSLA in Tamale, the Northern Regional capital. The trained volunteer then gave a step-down training to other members of the group. The community volunteers were given an intensive one-day training at the conference hall of the Christian Council Guest house in Tamale on the 17th of January 2018. They were taken through the concepts of the VSLA by Mr. Samuel Wangul, a facilitator from SEND-Ghana.

Participants were also given training on the drafting of the VSLA constitution and how to keep records on savings passbooks. A practical session was also conducted to assess participants' understanding of the concept and to identify challenges they will encounter during record keeping. They were also taken through the process of issuing loans to group members, how to calculate interest rates as well as recover loans from members. The team then presented a set of VSLA kit to each group to enable them to start the savings process in their respective communities.

Mrs. Dachia Midana and Hajia Poanaba Sumani, leaders in the Gbimsi *"Tilanngum"* VSLA, say that savings will be used to expand their farmland *"We want to use this VSLA savings to cultivate 60 acres of groundnut in 2018 cropping season."* The Gbimsi *"Tilanngum"* VSLA, located in the Gbimsi community of the West Mamprusi District of Northern Region, has a total membership of 30 volunteers who are all female. The VSLA has eight elected executives to manage financial affairs of the group and meets once a week to make weekly contributions. The group has engaged the service of a male secretary who helps them in their day-to-day record keeping, since all members can neither read nor write. The group has also been subdivided into five "Solidarity groups" with a membership of six per group for the community groundnut seed production.

In Gbimsi *"Tilanngum,"* as all other communities where VSLA were created, a training process to also sensitize chiefs and land owners at the community level on the need to release fertile arable lands to female farmers was largely agreed upon. *"This has given hope that women will have better access to arable land for groundnut and other crops cultivation,"* says Mr. Sardi Linus Handua, the male Secretary of the VSLA Gbimsi *"Tilanngum."*

The VSLA are also giving a relief to members who can contribute better the school fees for their children. According to Handua, the average school fees paid yearly for primary student is 1000 Ghana Cedi (about 200 US Dollars). *"I joined the VSLA because I can save money, get credit for income generating activities that will enable me to pay school fees for my 4 children who are all in senior high school,"* says Ms. Dachia Midana. *"Moreover, I can get money to prefinance my children school fees."*

## 8.3   Cutting Down the Breeding Cycle and Revamping the Groundnut Seed Systems

From 40 kg to 6 tons of breeder seed, 8 tons of foundation seed, and 37 tons of certified seed in 2017, Ghana is back to groundnut production

> *The fact that we can even mention an existing groundnut program in Ghana is in itself a success. Before the Tropical Legumes projects, there were no groundnut breeding program in Ghana,*—Dr. Richard Oteng-Frimpong reported.

Groundnut is the most important grain legumes in Ghana and CSIR-SARI is mandated to carry out the research on groundnut in Northern Ghana. *"Up to 2006, there were no funding to support the research on this crop, until Tropical Legumes started and were road- on in 2012,"* says Dr. Richard Oteng-Frimpong (Fig. 8.9), the groundnut breeder and seed systems scientist at SAR Ghana.

Richard first joined the projects with his research center in 2012 to start a groundnut program with 40 kg of available breeder seed, along with 57 new breeding lines obtained with ICRISAT as part of the Tropical Legumes projects. *"In the same year, we were able to conduct single location trial,"* says Richard. However, between 2013 and 2015, Richard says the program slowed down because there were no groundnut dedicated breeder.

**Fig. 8.9** From left to right: Dr. Richard Oteng-Frimpong, groundnut breeder, and Mr. El Hadj Abdul Razak, Director General of Heritage Seed Company Ltd. (Photo: Diama A)

In 2015, when Richard resumed work in SARI as a breeder and with the support of the Tropical Legumes, the Ghana groundnut breeding program eventually started again from scratch. *"We collected 300 new germplasm from the ICRISAT and since, we have been able to develop 109 new breeding lines out of them. Now we have our own breeding nursery and crossing program with 16 new crosses developed in 2017. Our target is to reach up to 20 new crosses per year."*

This breakthrough is also linked with increased on-farm participatory varietal selection (PVS), where up to 44 PVS were conducted in 2016 by an institution where groundnut breeding was nonexistent. Furthermore, the program has established a groundnut inspection plot to speed up the release of new varieties. *"This plot is established for inspection by the groundnut release committee. If it goes well, we are expecting to release 3 new varieties in 2018. Sensorial evaluation and nutritional varieties have also been completed as well and with regard to all aspects, the varietal release committee will make a decision,"* Oteng-Frimpong explains.

> *Tropical Legumes II and Tropical Legumes III projects have revived groundnut breeding program in Ghana,* confirms Dr. Roger Kanton, Deputy Director of CSIR-SARI. *We started from zero and now we have a fully fledged groundnut breeding program. We have breeding materials developed inhouse from our own crosses. With the new germplasm obtained from ICRISAT and other partners within the project, we were capacitated to undertake more experiments on yearly basis, and by so doing, we are cutting down the breeding cycle,* Kanton adds.

In 2016, SARI groundnut breeding program successfully produced 6 tons of breeder seed, 8 tons of foundation seed, and 37 tons of certified seed. *"No one will say that SARI groundnut program is dead as they use to say. Tropical Legumes and the USAID funded groundnut upscaling projects came to revive groundnut program in Ghana. Without any doubt, we would not have reached to this extent without the support of both projects,"* Oteng-Frimpong adds.

Partnerships were very instrumental in the success of the new groundnut breeding program. With the additonal support of the USAID-funded groundnut upscaling project, the TL project in Ghana extended and targeted more communities using participatory varietal demonstration plots and field days. *"There were not much varietal demonstrations before TL projects. The existing varieties were kept without promotion of any kind and only very few farmers knew about them. With the TL projects, 57 PVS were conducted in 2016 and up to 65 PVS in 2017."*

Also, we have acquired car to facilitate mobility. *"Due to our high mobility, we have moved from one multi-locational trial up to six multi-locational trials in a year. The multi-locational trials conducted in the Guinea Savanna only have been extended into the Soudan Savanna."*

With new irrigation facilities which allow to phenotype for drought tolerance traits in the existing germplasm, the groundnut breeding program can plan bigger for projects. *"We have been able to acquire a SPAD chlorophyll meter, a portable leaf area meter, a data longest that allow us to set up our own weather portable station and enable us to measure soil moisture level and also the environmental condition. All these facilities are helping us to develop drought tolerance materials for our target region. With the irrigation system functioning, we can have two cropping seasons in a year, we are able to produce early generation seed also on these facilities."*

Since 2015, the Tropical Legumes projects and the USAID-funded groundnut upscaling project have reached over eight thousand (8000) farmers (including 45% of women) and this is expected to continue increasing during the coming years. According to Dr. Oteng-Frimpong, this is because they have strong basis foundation to build upon. The TL projects which covered only three districts in 2015 is now targeting 30 districts in Ghana.

## 8.4    A Very Innovative Way of Making Foundation Seed Available to the Seed Producers

The increase in breeder seed production has improved the access of seed companies to foundation seed and out-growers are producing more certified seed. *"Four years ago, there were no breeder seed of groundnut available; we were all relying on one popular variety called Chinese,"* recalled Mr. El Hadj Abdul Razak (Fig. 8.10), Director General of Heritage Seed Company based in Tamale, Northern Ghana. *"In 2013, I heard of an improved variety but could not get access to it."* In 2017, the seed company produced about 60 metric tons of the improved groundnut Samnut 22 (shelled). The customers of the seed company include a network of 300 seed out-growers and individual farmers to whom small seed pack of 20 kg of the improved variety Samnut 22 have been distributed to.

*The project found a very innovative way of getting foundation seed into the seed value chain. Breeder seed problem is of the past and as a seed company, we have a big market for improved varieties of groundnut and cowpea,"* says Razak. *"Many farmers were equipped*

**Fig. 8.10** Left: Mr. Rachid, Research technician, and Dr. Richard Oteng-Frimpong, groundnut breeder, are happy to show breeder seed ready to be distributed to seed companies. Right: Mr. El Hadj Abdul Razak, Director General of Heritage Seed Company Ltd. (Photo: Diama A)

*with knowledge and skills in cowpea and groundnut production, and connected to our company with production contracts so that they can produced and injected enough certified seed. Before then, our company was very limited to only two communities. Now we work with seed out-growers in 8 communities.*

Mr. Fuseini Zaanyeya (Fig. 8.11) is member of one nucleus farmers' group involved with Heritage Seed Company in production of certified seed. He joined the network of the nucleus farmers of Heritage Seed in Gberimani-Tibogu community, Tolon District, Northern Ghana. *"Someone told me that they have a group supported by a seed company. The fellow introduced me to the Director General of Heritage Seed Company; that's how I came into cowpea production. He gave me 10 kg of cowpea seed which I first planted on 1 Acre and harvested about 0.5 tons in 2015. The following year, four other youth in the village embarked on cowpea production where we were all given 50kg by the same seed company. In 2017, we were 25 farmers using a total of 250 kg of the improved varieties provided by the seed company."*

Before then, like several other youth from his community, Mr. Fuseini Zaanyeya used to travel to the capital city of Accra in Southern Ghana where he sought in vain for greener pastures to better his standard of living. His dream to secure a tractor for expanding his field operations for increased productivity and better living standard came true when he returned to his community and settled on cowpea production as a business. He is currently the chairman of a farmer group engaged in cowpea seed business under Heritage Seed Company within the Gberimani-Tibogu community of Northern Ghana. Mr. Fuseini Zaanyeya says he can now afford to pay school fees and hospital bills for his children.

With his earnings, Mr. Fuseini Zaanyeya was able to purchase a new tractor which he now uses to render service to fellow farmers in the village. More recently he bought a motorbike which he says will enhance his mobility from his village to the district capital city (Tolon) where he has acquired a piece of land. He plans to construct a new house for renting and get money and settle a house for his two children in preparation for them getting into senior school.

**Fig. 8.11.** First from left: Mr. Fuseini Zaanyeya with his new motorbike. First from right: Mr. Fuseini Zaanyeya with members of his household. Down: Mr. Fuseini Zaanyeya standing in front of his tractor with his household members and some youth in the village, Ghana (Photo: Diama A)

Following Mr. Fuseini Zaanyeya example, many youths have come back to agriculture in Gberimani-Tibogu community. *"Many other youths are still carrying stuff in Accra capital city, but I am away ahead of them; they are suffering over there,"* says Fusseyni who is now his own employer and employs more than 25 workers. Mr. Fuseini Zaanyeya believes that more youth come to the village during cropping season but will migrate back to cities during dry season. He believes that a support in getting an irrigation facility could help settle more of the youth in the village.

Ms. Abibata Yiri (Fig. 8.12) is a member of one nucleus groundnut producers. After the cropping season, she works as a temporary staff of the Heritage Seed

**Fig. 8.12.**  Ms. Abibata Yiri (on right), a member of a Heritage Seed Company nucleus farmers in certified seed production, Ghana (Photo: Diama A)

Company. *"I am a groundnut seed out-grower to the company. I produce groundnut and sell it to the company. I get more profit when I sell my produce to the company as compared to the market. It has brought a big change in my life as I can contribute to school fees for my children including providing them with money for their lunch at school."*

After 12 years of activity, the Tropical Legumes (TL) projects come to an end the year 2019. In addition to quantitative impact assessment, this publication brings to light various stakeholders' own words about the kind of benefits they have made thanks to the TL projects. Research institutions, management and technical staff, extension services, NGOs, public and private seed companies, agro-dealers, farmer cooperatives, farmer groups, individual seed entrepreneurs, farmers, women farmers, processors, farm implement makers, middlemen, and traders have all shared their unique impact stories for being part of TL families.

From poor knowledge of recently developed legume technologies, most communities in the project areas of the target countries have been widely exposed and have a very good knowledge and use of newly released varieties of groundnut, common bean, cowpea and chickpea. This allowed them to increase crop productivity and production, improve resilience to advert weather conditions, and enhance family welfare through various livelihood assets such as ownership of land, houses, transport means, market connections, social recognition, food security, children education, health care, etc. The huge smiles on different faces pictured in this publication is illustrative.

Women conditions have tremendously improved as TL interventions specifically targeted women empowerment through capacity building and involvement in seed business. Many of the under-served women that the TL projects supported have had their living standards and those of their respective families in the target Eastern and West Africa countries lifted higher.

Community food security and resilience building is not a one initiative task. We strongly believe that future research and/or development interventions will leverage on the current dynamics set out and nurtured throughout the 12 years of TL projects. We are proud to deliver on this as a team made of diverse national and international partners.

© The Author(s) 2020
E. Akpo et al., *Sowing Legume Seeds, Reaping Cash*,
https://doi.org/10.1007/978-981-15-0845-5_9